Contents

Introduction

This book is aimed at undergraduate students of the applied social sciences (which may also include some aspects of human geography and economics). You may be studying a freestanding research methods module, or a module that requires the use of social research methods. In either case, this book is designed for you.

The assumed level of mathematics knowledge and attainment is no higher than GCSE or equivalent level. No knowledge of statistics and applied probability is assumed.

Each chapter focuses on incremental development of understanding of statistical principles. For readers who are less confident about their mathematical skills there is good news! Most applied statistical procedures exploit the use of IT to do all the hard work. The program most widely used in the social sciences is SPSS (Statistics Package for the Social Sciences). This is a commercial package that is also widely used outside the academic world. Opinion pollsters and market research organizations, for example, employ SPSS to analyze their data used to present their findings. So, SPSS is not just a tool used by academic statisticians, it is an industry standard tool and anyone looking to work in the field of commercial market and social research will need to be competent in the use of the package. This book will introduce you to the features of SPSS you are most likely to need, and develop your skills through exercises in each chapter. The approach is very simple; a chapter will:

- Explain the underpinning theoretical principles behind the employment of statistical techniques, and how to interpret the results gained; then,
- guide you through the process of using SPSS to apply the techniques using real data,

In most cases, you will not be asked to memorize, prove or use complex mathematical formula. The focus is on identifying appropriate techniques that you should employ and then use SPSS to do the calculations and, where appropriate, visual presentations of the data. The important thing to remember is that you are not required to be a mathematician. The skills of knowing how to collect the right data, understand what statistical techniques (and assumptions) to use, and understand what the results tell you are the province of the applied social scientist. SPSS is the tool you will use to help you.

Of course, since SPSS is an industry standard computer program, it is expensive to buy and licence. The good news is that if you are a student, an inexpensive educational licence will be available from your institution so you can have your own copy—although this may be for a limited period.

How to use the book

The book has been written so that statistical knowledge and skills are developed sequentially in a linear way. By that I mean that chapters are organized in a way that follows a logical order that reflects the need to understand one concept or skill before moving on. In that respect, it is not a reference work that you can easily dip into at any point. For example, you will need to understand the principles of probability (the likelihood of an event) before leaning about probability sampling (i.e. understand how and why samples are drawn from a research population).

The book should not be considered as a 'course' of instruction, but you will find it invaluable as a reference source if you are being formally taught research methods and statistics. The book will support that learning using language that is generally without jargon, and presents knowledge and skills using non-mathematical language—well, almost!

Each chapter is structured in the following way:

- A clear set of learning outcomes;
- Learning material set out to meet the learning outcomes;

- Summative activities for you to check learning against the learning outcomes for the chapter, with detailed feedback being made available from the Studymates web site. Prepared datasets in Excel and SPSS version 12 are also available from the website.

First principles: What's in a number?

Learning outcomes:

This chapter explores the nature of the use of *numbers* in statistics. By the end of this chapter you should:
- Understand the concept of 'numbers' in relation to statistical processes;
- Know that types of quantities will affect statistical methods applied to them;
- Understand the differences between and properties of nominal, ordinal, interval data and dichotomies, and their roles in statistical processes;
- Understand and be able to provide examples of discrete and continuous data and their roles in statistical processes.

What's in a number?

In the Introduction, it was explained that the purpose of this book was to help dispel some of the anxieties that numbers, or rather their meaning and manipulation, present for many people. In the context of this book, we are considering how numbers are used and manipulated in social sciences research (e.g. applied social research and psychology). It is essential that the meaning of a number is understood as this helps us design social science research, data collection and analysis.

Most of us take numbers for granted, although we may be less confident about using them in mathematical processes. There is something universal about them in a way that language is not. For example, the visual representation of numbers is more or less the same throughout the world, and their symbolism is certainly universal. Send a letter to a friend and there will be a number in the address that acts as a locator—a geographical marker (district in a city or house number). Buy souvenirs in Italy, and we will understand the value of what we need to pay. For athletes, numbers have particular positive or negative associations. It is important for them to come first, be the fastest, throw, hit or run the furthest. To come first is everything: last is nothing.

We are frequently identified by numbers rather than by our names. Some of these may be in *alphanumeric* format— a mixture of numbers and letters. Your National Insurance number is a typical example, and this uniquely identifies you in the UK population—no-one else has the same NI number as you (or at least they shouldn't have!). Your tax reference code is another. So while there may be more than one Ian Hosker in the UK, each of us will have our own unique *code* that identifies us for the purposes of tax, health and a wide range of other services provided by the state.

Numbers provide us with tools to communicate and work with the concepts of:

1. identification;
2. quantification;
3. ordering or position.

Properties of numbers and their use in statistics

The meaning of a number is very much dictated by the context in which it is used. Before looking at statistical processes and procedures, the properties of numbers need to be understood so that they may be used and manipulated correctly in quantitative social science research. Numbers fall within one of three main categories.

- Nominal (or identifying) numbers
- Ordinal (or ordering) numbers, and
- Interval or scalar (or quantifying) numbers.

Nominal

Numbers used for the purposes of identification are referred to as *nominal*. *Nominal* numbers cannot be used for counting or other forms of arithmetic operations (i.e. addition, subtraction, multiplication and division). Why not use letters or names, as this is what we are used to? In principle, no reason at all, but numbers are often more convenient and when using computer software to manipulate data, numbers are handled better by the program. For example, data entry is more likely to be error-free using numbers rather than text.

Nominal numbers provide a means of codifying a concept, and this is used extensively in statistical analysis, such as multiple option questions where each option is given its own code.

Consider the following question in a survey:

How did you travel to work today?

When you travelled to work this morning, which of the following methods of transport did you use? Write the number of your method of travel in the box below:

Option number

Method of travelling to work	Option
Train	1
Bus	2
Bicycle	3
Motorcycle	4
Your own car	5
As a passenger in someone else's car	6
On foot	7
Other	99

Table 1.1: Codes used to uniquely identify methods of travel

If Other, what is your way of getting to work

This approach of allocating a unique ID number to an option is more convenient than using text for several important reasons. Firstly, the person responding to the question will find it easier and quicker (and so less irritating) to write a number rather than the complete text. Secondly, computer programs are now used more or less universally to analyse research-based data. Numerical data entry is much easier and quicker and is more likely to be done with fewer errors than with lengthy text.

Note that in Table 1.1 there is an 'Other' category and this is coded 99. When you design a questionnaire, it is not always possible to think of all the possible responses, so a catch-all category may be used—and make sure you allow the opportunity to explain this. As you review questionnaire responses, you will be able to generate new categories and assign their own codes. The point about the 99 code for 'Other' is that it leaves enough codes to use for the new categories (i.e. 99 − 7 = 92 possible new categories!).

Anonymity of people responding to social surveys is an important ethical (and often legal) consideration. In fact, the *Data Protection Act 1998* and the *Freedom of Information Act 2000* both compel organizations collecting personal data to protect the interests of those from whom data has been collected—the *data subjects*. By allocating a unique identifying number to each respondent, the data may be processed statistically and the results published without revealing anyone's identity. If you are a student, then your institution will hold your details on a database and the record will not rely solely on your name. After all, if your name is Andy Smith that is not unique—a quick search on Yahoo! produced 44 379 hits. While some of these will be duplicates (i.e. references to the same person) it does illustrate the possibility of mistaken identification. Many actions in life do depend on the correct identification of an individual—so a unique numerical identifier assumes a critical importance in this role.

It should be self-evident from this explanation that if you attempt any mathematical operation on nominal numbers (i.e. + − × ÷), the result will be meaningless and even downright bizarre. To illustrate this, the box below shows the logical result of applying simple rules of number to a couple of values from **Table 1.1**:

> The logical result of applying the rule of multiplication to two of the number codes in Table 1.1
> $$2 \times 3 = 6$$
> actually means in the context of
> 'Bus' multiplied by 'Bicycle' equals 'As passenger in someone else's car'.

Clearly, this is nonsense, and that is the key to understanding what *nominal numbers* are about. They provide us with a means of easily identifying, or codifying, a unique property. The numbers themselves cannot be manipulated mathematically.

Dichotomies

Where the options in a category are in the form of 'either/or', i.e. one of two possibilities, the resulting coding is referred to as a *dichotomy*. Examples of dichotomies include: male/female; yes/no; true/false. The most common nominal numbers allotted to them are 0 and 1, or 1 and 2; e.g. yes = 1: no = 0.

Dichotomies are used extensively in statistical analysis. The important thing to remember about them is that the two number options used in a *dichotomy* are providing a unique way of differentiating between two alternatives that are often opposites. They are extremely important in statistical analysis because their use to differentiate between two distinctive groups within the research enables comparisons between the two. For example, gender represents how a *dichotomy* is used. By allocating a separate code for males and females (e.g. 1 and 2), a database can be split so that only data on males are analysed, or the differences in data between the two genders can be explored—such as comparing the salaries of men and women or their career progression, spatial awareness or ability to multi-task.

Ordinal numbers

Ordinal numbers provide a sense of ordering or position within a range of options. Look at the following example:

When you bought your last electrical appliance, consider the factors that influenced your choice. Place each of the following in your order of preference, from 1st to 5th:	
Factor affecting choice	**Order of importance**
• Value for money • Brand name • After sales service • Ease of use • Energy efficiency	

This information allows a marketing department to gain a view of the importance people place on the factors involved in decision-making. So, if you place 'Brand name' 1st and 'After sales service' as 5th, you cannot say that 'Brand name' is 5 times more important than 'After sales service'. Ordinal numbers only allow you say that 'Brand name' is the most important factor and 'After sales service' is least important to you.

In other words, *ordinal numbers* are not normally able to be manipulated mathematically as they only give a sense of position, or preference. There is a very limited set of circumstances where you can apply the four rules of number (i.e. $+ - \times \div$) and these will be discussed in Part 2 of the book when we consider the use of *inferential statistics*.

Even with that limitation on manipulation, *ordinal numbers* have very powerful analytical uses—such as in measuring people's attitudes, preferences, rating of service quality or experience.

Interval (scalar) numbers

These are quantifying numbers—e.g. size, volume, time interval, salary, height, weight, age, time. They are numbers that we use to count and carry out calculations. They actually represent quantities and can be manipulated by arithmetical processes. We understand that £1000 has double the purchasing power of £500, for example, and that by increasing £1000 by 10% there will be an additional £100. We can apply principles of ratio and proportion to interval numbers. In statistics, we may apply a wide range of powerful procedures.

The power of *interval* numbers can be literally illustrated using Fig. 1.1

Each turbine in Fig 1.1 is capable of producing a given amount of electrical power under any given set of conditions. Five turbines will clearly generate five times that amount of energy. This is an example of our general experience of numbers. While we are aware of *nominal* and *ordinal* numbers and their properties, we tend to associate numbers with 'maths'—i.e. capable of being used

Figure 1.1: The power of wind turbines

in mathematical operations that quantify our view of the world. *Interval* numbers, or rather the techniques by which they are manipulated as part of a problem-solving process, is the area that causes anxiety in many people, generating a significant issue of confidence in their use. This book should help to dispel much of that anxiety in the use of statistical procedures in the social sciences, where students may not have the advantage of studying maths or any of the 'hard' sciences past GCSE. Many students of the social sciences will not have a strong background in statistics, yet will be required to use statistical data analysis techniques in their coursework.

Numbers and statistical processes

Given that many people associate numbers with 'maths'—often in a fairly negative and anxious way—it is important to be very clear about how numbers will be used in any social science research. The next section on grouped data shows this very clearly. When a piece of research is being designed, particular attention needs to be paid not just to the research methods used, but also on how data will be manipulated as part of the analysis. In other words, a key element of experimental design is the decisions made about the statistical techniques that will be employed.

This is an oversight failure by inexperienced researchers. Too little account is taken of why data should be collected in a given format—when is it appropriate to record information as *nominal, ordinal* or *interval*, quite apart from how should a questionnaire be designed to gather the data in the format needed.

The decisions are determined by the nature of the property you are recording and will be discussed in more detail elsewhere in the book. For now, it is sufficient (i.e. essential) that you understand that numbers—like our language—may have multiple meanings determined by the context in which they are used.

Continuous, discrete and grouped interval data

Interval data forms the backbone of statistical analysis, allowing us to interpret and quantify statistical relationships and their significance. There are some additional properties of *interval* data that need to be understood so that you can handle data efficiently and effectively.

Continuous data

Some forms of *interval* data have no minimum size of unit. Length is a good example of a *continuous interval* data property. For example, a piece of string may be 10.5 cm long, or 10.6 cm, or anything in between, such as 10.56736519887 … . The only limitation placed on quantifying length is in the accuracy with which we are able to measure it. When using continuous data, it is good practice to indicate the level of accuracy we are using (e.g. 'to the nearest millimetre').

Discrete data

Currency is a good example of discrete interval data in that there is a minimum size of unit involved. In the UK, this is the penny, and with the Euro, it is the cent. Theoretically, the money markets tend to ignore this and often express exchange rates in a form that simulates a continuous measure—e.g. £:€ = 1.385 (i.e. specifies fractions of a cent). However, when actually converting £ to €, the final figure will be rounded to the nearest cent (as there are no coins smaller than a cent).

Grouped data

There will be occasions when you will either collect, or need to convert interval data values into groups with upper and lower limits. For example, you may ask people in a survey to identify an age group or salary range that accurately represents their circumstances:

25–34
35–44
45–54 etc.

or

Under £10,000
£10000–£19,999
£20,000–£29,999
£30,000–£39,999
£40,000–£49,999
£50,000+

A key point to remember here is where the category threshold values lie, and this is much easier when you are dealing with discrete data values. The age categories are based on whole year units, while salary levels in this example are based on £1 units. If you use this approach, you will need to be clear about where boundaries lie. Avoid creating category lists like:

20–30
30–40
40–50

The problem here is that if you are 30, which category do you use? In other words, you need to identify the category boundaries so that two conditions are met:

1. The categories do not overlap at the boundaries and so cause confusion;
2. The boundary limits must be set so that no-one falls between the limits (i.e. they are neither in one or the

other adjacent category—or can consider themselves to be in such a position).

The rules on boundaries between categories need to be made very clear to the person responding to the question being asked

Category properties—collecting grouped or ungrouped data?

Notice that each of the data set categories above, with the exception of the salary group £50,000+, are of equal size range—e.g. in the salary example, each group spans £10,000. There are good reasons for doing this as we shall see later when looking at statistical techniques. Groupings of equal sizes make working with the data much easier. However, notice also that the final salary group is something of a 'catchall'. The number of people earning £50,000+ is relatively small so when collecting data like this from a survey it is also useful to ask for the actual figure as this allows you to create new categories later—e.g.

£50,000–£59,999
.... and so on.

This will also make it easier to manage the statistical analysis.

As you may not be absolutely sure what people's salaries are going to be when you undertake a survey, the catchall is useful as it avoids the difficulty of developing a long list of categories on a questionnaire, as it is perfectly possible for you to survey someone earning £500,000 pa!

Using categories of equal sized intervals is not obligatory, and there are occasions when data is grouped in categories of different sized intervals. You need to be very clear about:

- Why you are choosing to present grouped data;
- Why you have selected the group boundaries.

Figure 1.2 is an example of an effective visual presentation of grouped age data derived from the 2001 census. The interesting point to note here is that the groupings are not all the same size and several have considerably larger intervals than the others.

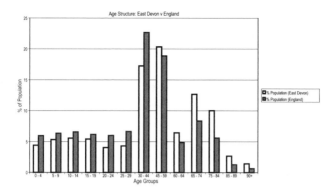

Figure 1.2: Using grouped data for visual comparison (*Source*: ONS–2001 census)

Grouping data like this for visual presentation allows the pattern of distribution to be seen more easily. Note that most of the age group intervals are the same—i.e. 5 years—and the three groups that don't conform to that structure are significant in terms of social policy. This chart was created using the grouped data published by the Office for National Statistics (ONS), and this obviously matters to government. For example, the two large groups of 30–44 and 45–59 are significant in that taxation (by central and local government) raises greater contributions per capita from those in these age groups. On the other hand, the greatest demand for public services generally comes from the younger and older age groups.

Grouping data for visual presentation in this way allows the viewer to see the obvious comparisons between East Devon and England. For example, East Devon is considered to have a very high proportion of older people. So what does this chart tell us:

- Relative to population size, East Devon has an age structure in which the proportion over the age of 45 is higher than that in England as a whole;
- This trend increases with increasing age.

While this apparently confirms the view that East Devon is a 'retirement area' and would suggest that the demand for services for older people will be proportionately greater here, the way the data is presented provides only a crude and limited capacity for analysis.

As far as statistical manipulation is concerned there are some problems associated with the use of grouped data. Wherever possible collect data in a disaggregated format (e.g. ask the question: How old were you at your last birthday?) as this absolute data can be more easily manipulated. However, information such as age and salaries are quite sensitive, and people are not always willing to reveal absolute values, but may be more amenable to placing themselves within a broad range of values. The counter-argument is that by collecting data at its most disaggregated level, the researcher is given greater opportunity to interrogate, evaluate and analyse it to develop models (hypotheses) of what is going on.

Exercises

Activity 1: Your personal numbers

Copy the grid below and list your own important numbers that fall within each conceptual category.

Identifying numbers (nominal)	Quantifying numbers (interval/scale)	Ordering numbers (ordinal)
e.g. parking space number	e.g. bank balance	e.g. batting order of your cricket team

Activity 2: Practical examples of grouped data use

This activity asks you to identify examples of where grouped data is used to make data presentation and interpretation easier. Look through the textbooks used on your course, the Internet, newspapers, magazines and journals where statistical data is being presented and:

1. Identify examples of grouped data presentation.
2. For each example attempt to explain why the writer/ researcher has chosen to present the information in this way.
3. What details of the information may have been sacrificed to make it visually better presented to show relationships?

2 ▮ Managing your data—the use of statistics software

Learning outcomes:

Research in the social sciences involves gathering a great deal of data which has to be managed. The use of statistical analysis is not just about maths, it is very dependent on the systematic collection and organization of data. The purpose of this chapter is to introduce you to the main professional software package used in the social sciences–SPSS. Both packages are available at considerably reduced cost (perhaps even at no cost) if you are eligible for an educational licence. We will also explore some of the features of the software because most of the statistical processes explained in this book can be carried out with the aid of these computer programs. By the end of the chapter, you should:

- Understand the need to use a systematic approach to recording data;
- Know how to use SPSS for recording and managing statistical data;
- Be able to enter data into a statistics software package and explore their characteristics;
- Be able to use and interpret a 'coding frame';
- Be able to collect and store data in SPSS.

Why use computers?

We can probably all recall a time in our mathematics education when we have been banned from using an electronic calculator to find the answer to a question. The rationale for this, quite rightly, is that we need to learn the fundamental skills of calculating using appropriate techniques. The approach taken in this book is that the techniques will be described and explained so you gain an understanding of what is going on, but the hard work will be taken up by appropriate computer programs. This represents the reality of contemporary professional social science practice. All the techniques described in the book can be carried out by hand—and you will be shown how. While the ability to do so is an important skill, perhaps more important is the ability to identify and apply the correct techniques and procedures. Computer software takes out the hard work and produces results more quickly than can be done by hand.

The industry standard program in common use by both academic and commercial applied social and psychological research is SPSS (Statistics Package for the Social Sciences) and the use of this will be taught through practical exercises and projects.

Systematic data collection and recording

Later chapters will teach the skills of research design, including questionnaire design and one of the key issues to be aware of is the need for a systematic approach to identifying the nature of the data needed, and the systematic approach to collecting and recording it for later analysis. Get this wrong and your work as a researcher becomes incredibly difficult.

SPSS tutorial

If you are familiar with the layout of spreadsheets—i.e. cells in rows and columns—and have used them to develop small databases, then the look of SPSS will be familiar to you. In this section you are asked to take a short tour around the interface of each package, and familiarise yourself with some of the basic features. The more advanced features will be examined as they are needed in later chapters, but it is important at this stage to get a feel for the software.

The following short tutorial will help you become familiar with the SPSS interface.

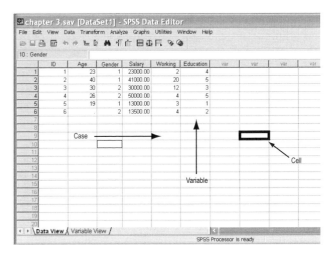

Figure 2.1: The data entry window of SPSS

Fig 2.1 shows the beginnings of a set of data that may have been collected to explore the possible relationships between employees' ages, gender, the total number of years in the labour market and current salary levels.

Fig 2.1 shows the Data View window that allows you to enter data. Individual data items are entered into *cells* and are grouped into *variables* and *cases*.

Variables

This is the characteristic or property that is being measured. In Fig 2.1, age is a variable as are gender, level of education and length of working life. Each column stores the individual data items collected on that variable. Note that each variable is given a name, and this will be explained in more detail below.

Cases

These are the subjects from whom data is being collected. In this example, each employee is a *case*. So far, data has been entered for six cases (employees). One of the variables has been labelled ID and the purpose of this is to be able to identify each individual case during analysis later, should the need arise. A good example of this is where the data seems odd or inaccurate, and this may be the result of an error in collecting the data (e.g. the age may be 23, and working life recorded as 30 rather than 3!). You can check it.

The use of a unique identifier is very commonly used even where the person the data relates to is unknown (e.g. as part of anonymous surveys). Questionnaires are usually coded and so if this information is included as an ID variable, you can go back to the original questionnaire and check for errors or anomalies. This is also particularly useful when checking *missing values*—see below.

In row (case) 6 there is '.' and not a number. This is a *missing value*. If data is not entered into a *cell* the program assumes the value to be missing. There are two likely reasons for this:

- The information may not have been provided or recorded when the data was being collected (e.g. the person may have refused to give the information).
- The person entering the data into the SPSS database may have missed entering the information—i.e. it may be a *data entry error*.

As there is a identifier for this case (i.e. the ID variable), the researcher can go back to the original data source, such as a questionnaire, and check.

Data entry and definitions

Data is entered into the SPSS data entry window directly from questionnaires and the way a questionnaire is structured to help you do this is explained in Chapter 5. One of the key points to keep in mind is that in designing the way you collect and record data, you have to consider the relationship between the structure of the questionnaire and the SPSS database that will store and analyse the data. In this chapter, the way you define the variables is looked at in some detail. Spending time on this is a good investment as it eases the process of interpreting the outputs of analysis later.

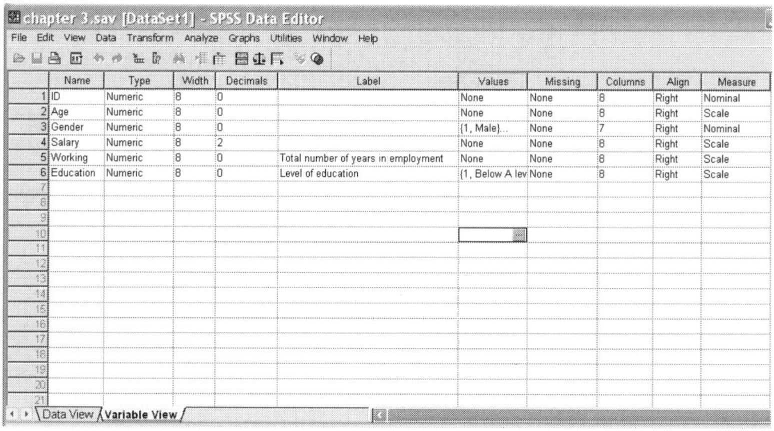

	Name	Type	Width	Decimals	Label	Values	Missing	Columns	Align	Measure
1	ID	Numeric	8	0		None	None	8	Right	Nominal
2	Age	Numeric	8	0		None	None	8	Right	Scale
3	Gender	Numeric	8	0		{1, Male}...	None	7	Right	Nominal
4	Salary	Numeric	8	2		None	None	8	Right	Scale
5	Working	Numeric	8	0	Total number of years in employment	None	None	8	Right	Scale
6	Education	Numeric	8	0	Level of education	{1, Below A lev	None	8	Right	Scale

Figure 2.2: The variable description window

At the bottom of the SPSS window there are two 'tabs', clicking the computer mouse on each 'tab' will toggle between the Data View (Fig 2.1) and Variable View (Fig 2.2).

This window allows you to define the variables used in your database.

Name

In this column you give each variable a name. Note that you can only use a single word not a phrase or sentence. For variables such as age and gender this is not a problem as they have an obvious meaning, but describing 'length of working life' or 'level of education' becomes more challenging. In this database, the variables have been given the names of 'Working' and 'Education' respectively, so to clarify the meaning of the variable, descriptive text is typed into the *Label* column. This text will be seen in the analysis printouts. The *Label* facility is extremely useful in reminding you of what the variable means.

Type

Here you are able to define the type of data that will be entered into this variable. Clicking on the Type cell opens a dialog box to select the data type—see Fig 2.3.

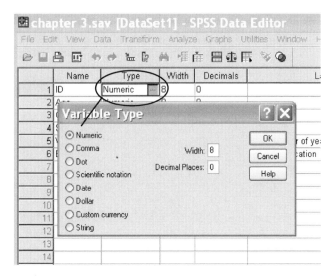

Figure 2.3: Data type dialog box

The most common formats you will use are numeric (number), date, and string (text rather than number). In this case, the ID variable is a number and you can set the properties of number of digits and decimal places by typing in values. These can be changed later if you need to.

Values

Chapter 1 introduced you to the three 'types' of number (nominal, ordinal and interval) and it was suggested there that questionnaire responses are best coded with a number as this makes it easier to enter the data into a database for analysis—for both speed of entry and accuracy. The *values* property of a variable allows you to define the meaning of each code. During analysis of the data, outputs such as tables and charts then show the definition not just the number and this makes it easier to interpret the information. Fig 2.4 shows the dialog box that opens when you click the mouse in a values cell. What you see in the illustration is the last code definition being entered. Once you have entered a **Value** and a **Label**, click **Add** to confirm it.

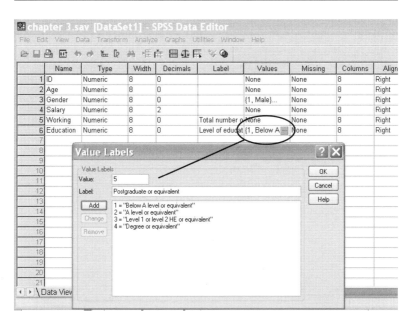

Figure 2.4: Assigning values to definitions

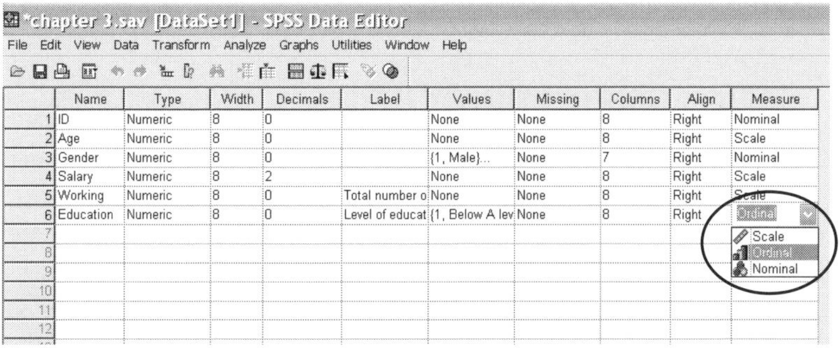

Figure 2.5: Selecting the type of measure used in the variable

Measure

This refers to the type of number: nominal, ordinal or interval (referred to as *Scale* in SPSS). Clicking on a cell in this column opens a drop down menu from which the form of number is selected (see Fig 2.5).

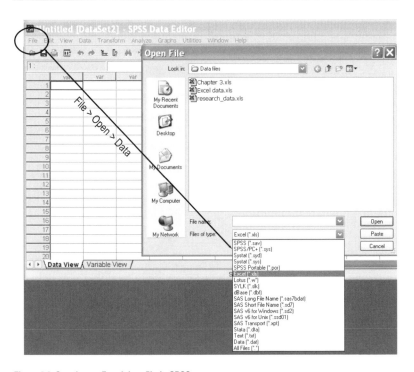

Figure 2.6: Opening an Excel data file in SPSS

Importing data files

SPSS has the very useful facility of being able to import data from other computer programs. For example, data may be stored in a database such as Microsoft Access or Excel and this data can be exported in a format that SPSS can read. This allows data to be made portable between computers and programs.

Fig 2.6 shows the process of importing data from another source. From the menu bar, select **File** > **Open** > **Data**. The **Open File** dialog box allows you to read a wide range of data formats (see the drop down menu in Fig 2.6). In this case, Excel files have been identified.

Understanding 'coding frames'

The problem with working with a database like the one used so far is that the data is in the form of numbers.

Each number clearly has a meaning (we discussed this in chapter 1) but unless this is written down for people using the data, the meaning will be lost. A good example of this occurred some years ago when a particular national survey was redesigned with a change of managing agency. There was some concern about the ability to make comparisons over time if the new data could not be compared with the old. It was agreed that 20+ years of very valuable data would be transferred to the new agency, and the database was duly transferred. There was a problem though in that the original definitions of the values in each variable were 'lost'. In other words, the data was rendered useless because they could not be interpreted!

The written document that defines the codes used in each variable is called a *coding frame* and we will come across this again in the chapter on research and questionnaire design (chapter 5). **Table 2.1** is the coding frame for the data used in this chapter to illustrate the use of SPSS.

Question	Variable in database	Coding frame
1. What was your age at your last birthday?	Age	Use the respondent's answer as written
2. Are you male or female?	Gender	1 = Male 2 = female
3. What is your current annual gross salary (rounded to the nearest £100)	Salary	Use the respondent's answer as written
4. In what year did you leave full-time education and begin employment?	Working	Calculate the number of years in employment from the date provided by the respondent

Table 2.1:
A coding frame

(Continued)

Table 2.1:
(Continued)

Question	Variable in database	Coding frame
5. What is you highest level qualification you hold?	Education	1 = Below A level or equivalent 2 = A level or equivalent 3 = Level 1 or level 2 HE or equivalent 4 = First degree or equivalent 5 = Postgraduate or equivalent

The *coding frame* makes the link between the question, the variable in the database and the response codes that need to be entered into the database. In other words, each response given by the *respondent* (i.e. the person answering the questionnaire) is coded prior to entry. In small scale research projects this will be marked on the completed questionnaire itself and the codes then transferred from the questionnaire to the database by the researcher or an assistant.

You will be doing a great deal of this as you work through this book!

Exercises

1. This exercise helps you to set up a small SPSS database from scratch. Later chapters will be looking at how to design your research or questionnaire, but don't worry about that for this exercise which is a simple task to collect data and create a database. You are going to gather data on people's reading habits—i.e. find out how many books people estimate as reading in a year—and see if there is any relationship between age, gender and the number of books read.

   a. Create a coding frame for the variables: age, gender, and number of books read in a year.

   b. Create the database in SPSS using the coding frame as a guide.

   c. Collect data by asking as many people as possible—see if you can reach a least 50, preferably more (why do you think larger numbers are better?). You could use a simple table like the one below to help you collect the data:

ID	Age	Gender	Number of books read in the last 12 months
1			
2			
3			
4			
5			

   d. Once you have completed the survey, enter the data into your database. **Don't forget to save the database by selecting FILE > SAVE on the menu bar of the program at regular intervals to avoid the risk of losing data.**

   e. Check your database against the data collected to make sure you have not made any *data entry errors*.

Correct any you find. You will use this database in other exercises later in the book.

2. This exercise asks you to import an Excel data file (satisfaction.xls) into SPSS, and use the variable labels facility of the program to label each variable in accordance with the coding frame supplied (coding_frame.pdf).

   a. Obtain the following two files from Studymates and save them to a convenient location on your computer:

      i. satisfaction.xls

      ii. coding_frame.pdf

   b. You will need to print the file **coding_frame.pdf** for future reference (and for this you will need Acrobat Reader on your computer—this can be downloaded from **_www.adobe.com_** if you don't have it).

   c. Open the data file, **satisfaction.xls** in SPSS. Select FILE > DATA from the menu bar and locate and select the Excel file on your computer. When you select OPEN, an OPENING FILE OPTIONS dialog box will appear. Tick the READ VARIABLE NAMES checkbox and click OK.

   d. There should be 246 records. Use the coding frame to label the variables.

   e. Save the file in the SPSS format, depending on which program you are using. We will return to this database in other chapters.

3 **Probability—
the principles**

Learning outcomes:

The theory of probability underpins the conclusions we make about data collected in social scientific research. A thorough understanding of the theory of probability explained in this chapter is essential if you are to understand the use of descriptive and inferential statistics explained in later chapters. It is difficult to over-estimate the importance of the principles of probability as the pre-requisite to understanding the use of statistics in all aspects of social science research and data analysis. The concepts introduced and explained in this chapter underpin the rest of this book and you may find it useful to return to this chapter from time to time to refresh your understanding.

Use this chapter to:

- Understand the underlying principle of probability;
- Be able to calculate the probability of success and failure of single events;
- Be able to calculate the probability of success and failure in multiple events which are:
  - Independent of; and,
  - dependent on each other;
- Be able to understand and use the convention of recording probability and the meaning of 0 and 1, including the sum of all alternative events being equal to 1;
- Understand that probability may be used to identify potential bias and is a tool for testing hypotheses through empirically derived data, i.e. *statistical significance*.

Probably or probability?

Probably

In an everyday, non-mathematical context, you always make judgements about the likelihood of a particular event. Where a number of possible outcomes exist, you assess which is the most likely under any given set of circumstances. For example, when I travel to work each Monday morning, I notice that the traffic is usually heavier, and the journey correspondingly slower than it is at the same time Tuesday to Friday!.

The judgement that is being made here is based on my experience of driving to work along a route that I have used for many years. I believe that on balance this is more *likely* to happen than not because it has been the case *in the past*. I am using historical evidence to predict a future event and assume this event will *probably* happen.

Probably is an intuitive assessment that all sentient animals make in relation to everyday existence—part of the survival kit—and the process becomes fine-tuned by exposure to similar events over time. In other words, I make a risk assessment about being late, so decide to begin travelling a little earlier on Mondays. Of course, I may get it wrong on some Mondays because school holidays result in less traffic along the route I normally take, which means I have time for two cups of coffee at the other end. My prediction has been *confounded* by another factor.

This seemingly trivial example does serve to warn us about the basis of our predictions—i.e. our interpretation of *cause and effect*.

- *Likely* is not the same as *will happen*—there is still an element of *uncertainty*.
- Assuming a direct causal relationship between a single factor and an event (in this case Monday leading to heavy traffic) fails to take account of other factors that may be at work and we may or may not be aware of these—we will return to this these later in Part 2 of the book.

Probability (or the mathematics of uncertainty!)

Probability looks at an event and attempts to represent its likelihood mathematically and so we begin to make statements such as: "There is a 20% chance that it will rain today". This requires us to identify factors that contribute to the event and to the possible outcomes. We know that if we spin a coin it has to land on one of two sides; there are two possible outcomes. But, is there a third possible outcome—landing on its edge? Experience tells us 'no', but "what if" we spin the coin in the air above soft ground? It could land on its edge and be held upright in that position.

Introducing another factor, or *variable*, alters the number of possible outcomes. Our assumption of there being two outcomes is now *confounded* by this third variable which suggests the possibility of a third outcome. The chance of any particular outcome occurring during any single event is reduced as the number of possible outcomes increases. The application of the theory of probability seeks to calculate the likelihood of a particular outcome. As we will explore later in the book, we also use probability to assess whether or not the outcome of an event is so unlikely, it is more likely that something important is going on that influences events—a causal effect that means the outcome is not the result of *chance*. We refer to this as *statistical significance* and this will be discussed towards the end of this chapter.

The likelihood of an event— calculating its *probability*

The implication is that there must be a mathematical route to arriving at such a conclusion and you will explore the route, and the underlying assumptions, in the remainder of this chapter. The simplest way of explaining probability is by using the example of tossing a coin and the likelihood of it resulting in a *heads* or a *tails*. This will be extended by looking at what is likely to happen when you throw a die. In that case there are six possible outcomes.

Spinning a coin

All things being equal, a coin is just as likely to land on heads as it is on tails. There are two possible outcomes, and each alternative is as likely as the other. The probability of it being heads is therefore said to be 0.5, with the same probability for tails. The probability is usually expressed as a decimal or a percentage (i.e. 0.5 or a 50% chance).

The total probability of all the possible outcomes adds up to 1 (i.e. 100%). Things are never that simple, of course, because if it lands on heads after the first spin, it does not necessarily land on tails after the second spin. This is because each spin is a separate event and the probability of landing on heads is still 0.5—so it is just as likely to land on heads again. **Table 3.1** shows the outcomes of 500 spins of a coin, listed in columns of 50 spins.

Table 3.1:
Outcomes of
500 spins of a
coin

1st 50	2nd 50	3rd 50	4th 50	5th 50	6th 50	7th 50	8th 50	9th 50	10th 50
H	T	T	H	T	H	H	H	T	H
T	T	T	T	H	H	T	T	T	T
H	H	H	H	T	H	T	T	H	T
H	T	T	T	H	T	H	H	H	H
H	T	H	H	T	T	T	H	H	H
T	H	T	T	H	H	H	H	T	T
T	T	H	H	H	T	T	T	T	T
T	T	T	T	H	T	H	H	H	T
T	H	H	H	T	H	T	T	T	T
T	H	T	H	H	T	T	H	T	T
T	H	T	T	H	H	H	T	T	T
T	T	T	H	H	T	H	T	H	H
T	H	H	T	T	T	T	H	T	T
T	H	H	T	H	H	T	T	T	T
T	H	H	H	T	H	T	H	H	T
T	T	T	T	T	T	H	T	H	H
H	H	T	T	T	T	T	H	T	T

(Continued)

1st 50	2nd 50	3rd 50	4th 50	5th 50	6th 50	7th 50	8th 50	9th 50	10th 50
H	H	H	T	T	T	T	T	H	T
H	H	H	T	H	T	T	H	H	H
H	H	T	H	H	T	T	H	H	T
H	H	T	H	T	H	H	T	T	H
H	H	T	H	H	H	H	H	T	T
T	T	T	H	H	H	T	H	H	H
T	H	H	T	H	H	T	H	H	T
H	H	H	H	T	H	T	H	H	T
T	H	T	H	H	T	H	H	H	H
T	T	H	H	T	T	H	H	T	T
T	H	T	H	H	T	T	H	T	T
H	H	H	T	T	H	H	H	H	H
T	H	H	H	H	T	H	H	H	T
T	H	H	H	T	T	T	T	H	T
H	T	T	H	H	T	H	H	T	H
T	T	T	T	H	H	T	H	H	T
T	H	T	H	T	T	T	T	T	T
T	H	T	T	T	T	H	T	H	T
T	T	T	T	H	T	H	T	H	H
T	T	T	T	H	T	H	T	H	H
T	H	T	H	H	H	T	H	T	H
H	T	H	H	H	H	T	H	H	H
H	H	H	H	H	H	H	H	T	H
H	T	T	T	T	T	H	T	H	H
H	T	T	H	H	H	T	H	T	H
H	T	T	T	H	T	H	T	H	H
H	H	H	H	T	T	H	H	H	T
T	H	H	T	H	T	T	H	H	T
H	T	T	T	H	H	T	T	H	T
H	H	T	H	T	T	T	T	H	T
H	T	H	T	T	H	H	H	T	T
H	T	H	H	T	H	H	T	H	T
H	T	H	T	H	H	T	T	T	H

Table 3.1: (Continued)

The table shows the results of each spin in consecutive order in the columns (labelled 1st 50, etc).

If the probability of each of the two possible outcomes of spinning a coin is 0.5, or 50%, then you might expect the table to be made up of 50% heads and 50% tails. In fact—and this is not a fix—a coin was spun 500 times to produce this table and the actual result is very close to that expectation. 50.4% of outcomes were heads (253 occurrences), and 49.6% of outcomes were tails (247 occurrences).

The basic method of calculating probability

Calculating the probability of an event's outcome is deceptively simple. The deception lies in the assumptions that are made, but this will be explained later. To calculate the probability of any particular outcome, you need to first determine the number of possible outcomes. The probability is then a fraction of 1.

Spinning a coin

In the case of a coin, there are only two possible outcomes—heads or tails—so the probability of each of these outcomes is 0.5 (50%) as was described earlier. The process can be represented visually with a 'tree diagram' and this is often a good way to show all the possible outcomes of an event prior to calculating the probability (Fig. 3.1).

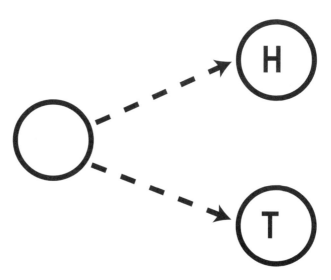

Figure 3.1: Tree diagram to show the possible outcomes of a single spin of a coin

Using mathematical terminology and symbols, we represent the probability of landing on 'heads' or landing on 'tails', or landing on 'heads' OR 'tails' as:

$$p_{(head)} = 0.5 = 50\%$$
$$p_{(tail)} = 0.5 = 50\%$$
$$p_{(head\ or\ tail)} = 0.5 + 0.5 = 1 = 100\%$$

. ... where **p** is a symbol to represent probability.

The probability is sometimes also expressed as a ratio—i.e. **1 in 2**—but it is more usual to present probability as a decimal or percentage. The ratio form can be useful to communicate the idea of a probability to the lay reader or where the decimal and percentage forms are awkward and inexact, as would be the case with throwing a die (see below) where the probability of each outcome results in a recurring decimal.

Since the coin has to land on 'heads' or 'tails' the probability of it being one or the other is 1 or 100%. In other

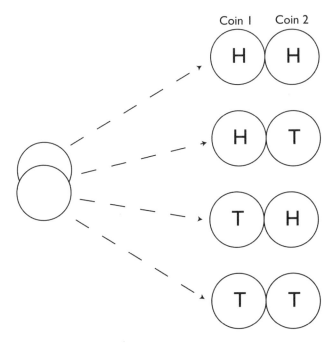

Figure 3.2: Possible outcomes of spinning 2 coins

words, the sum of the probabilities of all possible outcomes will always be 1 or 100%.

> **Q:** What is the probability of spinning two coins and both landing on heads?

Fig 3.2 is the tree diagram for the possible outcomes of that event.

At first sight you may think that there are three options: HH, TT, HT, but each coin is a separate entity so the options of HT and TH, while superficially the same, actually are different outcomes. There are four possible outcomes in that case. The probability of each outcome is 0.25 (25% or 1 in 4).

Throwing a die

A die has six faces, and hence six possible outcomes. Therefore, the probability of throwing a six is:

$$1 \div 6 = 0.166666 \text{ recurring } (0.17 \text{ if we round up}) \text{ or } 16.67\% \text{ or } 1 \text{ in } 6$$

The probability of throwing any single number will also be 16.67% as well. So the total probability of throwing any one of the six numbers is also 1 or 100%. Fig 3.3 is the probability tree for this 'event'.

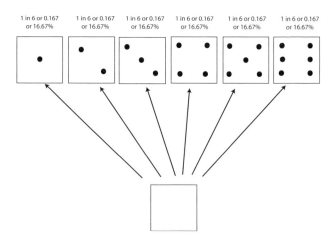

Figure 3.3: Tree diagram showing possible outcomes of throwing a die

The most conspicuous property of probability is that it is a straightforward proportion calculation, with the key point being that the probability of any single outcome is the same value as its proportion of possible outcomes.

'Success' and 'failure'

If a die is thrown then it can land on one of six numbers. If we want it to land on a particular number, say 1, and it does so, then we have a successful outcome. Obviously, it could land on any one of the five other numbers, and that would be a failure. This is the principle that lies behind gambling, of course.

It follows that;

$$p_{(success)} + p_{(failure)} = 1$$

Looking again at the die, the probability of it landing on one is 1 in 6 (16.67%), as we saw earlier. It is more likely not to land on one, and the probability of this is 5 in 6 (83.33%). In other words, you are more likely to fail than succeed.

$$p_{(land\ on\ 1)} + p_{(not\ land\ on\ 1)} = 0.167 + 0.833 = 1 = 100\%$$

The assumption of 'fairness'—the idea of 'bias'

There is an underlying assumption behind what has been written so far in this chapter, and that is we assume the calculated probabilities are based on *random* events that are the subject of *chance*. So, we are assuming that every number on the die has the same chance of occurring and there are no external intervening factors that distort this. In other words, the die is not 'loaded'.

This does present some difficulties. Throwing a die six times does not mean that each number will come up in turn because each throw is independent of those that preceded it. Every time it is thrown it was explained earlier that each number still has the same chance of occurring. You may have your suspicions if the die is thrown 100 times and 6 comes up 100 times! Is the die *biased*—i.e. loaded in such a way that it has been deliberately manufactured with an unequal weight distribution that pre-disposes it to land on six? Short of 'testing' the die, you cannot definitely say so because no matter how unlikely the result, probability theory tells you that if it is a possible outcome of chance then it could happen. The best you can do is voice your suspicion of *bias* and get it checked out because this outcome is so unlikely because of the very small probability—but it could happen by chance alone.

This demonstrates a clear example of how probability is used in the sciences—both natural and social. It alerts us to any result that seems very unlikely, on the basis of probability, to have happened by chance. The rest of this book will explore this in more detail, but in the meantime hold in your mind the idea that because a result is so unlikely to have happened by chance you cannot automatically assume bias—i.e. an external factor which is over-riding chance. If that is the case, can you use probability to predict the outcome of a future event?

Predicting a future event?

You know the probability of either a heads or a tails when you spin a coin, but what you will *never* know is what the outcome will be ahead of actually spinning it. You can assess risk of failure (i.e. 50% chance of not getting a heads next time) and on the basis of that knowledge make a decision about a bet. If the last spin was a tails, there is no guarantee that the next will be a tails. The two spins are completely independent of each other. You can see this in **Table 3.1** where, in the first column spins 6 to 16 produced a remarkable run on tails. This could not have been predicted.

> You should be clear that probability cannot provide a definitive mechanism for predicting an outcome. Probability does help you assess the likelihood of an outcome, and therefore informs your decision-making process. If the probability of an outcome is high, then you may think it is probably worth taking a risk with a particular decision.

Later in this chapter you will look at something called *statistical significance*. What this means is that you will use probability theory to assess whether or not an outcome is really the result of some factor under investigation, or is more likely to have happened by chance or random act.

The 'addition' and 'multiplication' rules

So far, we have looked at the probability of the outcome of a single event, such as the throwing of one or more die, or the flipping of a coin, but often we are concerned with the probability of a more complex set of possible outcomes. We need to understand the relationship between outcomes in order to accurately calculate probabilities. There are two basic rules we apply: the *addition rule* and the *multiplication rule*. The choice depends on our analysis of the problem. In this section, we will look at the calculation rules in more depth.

The addition rule

When possible outcomes are mutually exclusive

You throw a die and want it to land on 6. It cannot land on any of the other numbers as well. This makes each possible outcome *mutually exclusive.*

Since there are six possible outcomes, the probability is approximately 0.17 (i.e. 1 in 6). But what is the probability of throwing either a 1 or a 6. Since on a single throw it is not possible to have both 1 and 6 as an outcome (it has to be one or the other), then these outcomes are *mutually exclusive.* Each has an equal chance of occurring. The probability is calculated by adding the individual probabilities, i.e.

$$p_{(1 \text{ or } 6)} = p_{(1)} + p_{(6)} = 1/6 + 1/6 = 1/3 = 0.33 = 33.3\%$$

Gaining a 1 or a 6 represents two of the six possible outcomes—a 1 in 3 chance—as seen in the tree diagram Fig 3.3. The corresponding probability of not throwing a 1 or a 6 will be the probability of failure, i.e. 0.67—or a 2 in 3 chance.

In these circumstances where the outcomes are each mutually exclusive, the probability of gaining any one of two or more possible outcomes is calculated by adding the probability of each possible outcome

Where possible outcomes are not mutually exclusive

Now look at a slightly different example where the range of possible outcomes is more complex. The example used to illustrate this set of circumstances is a deck of playing cards.

Q: What is the probability of drawing either a spade or a king when picking a card from the pack?

Begin by assuming that the pack has been thoroughly shuffled and that the position of any card is not known, and that the selection of the card from the pack will be a random act—i.e. we have set up the conditions to ensure there is no bias.

It is often useful to represent this kind of problem visually using a *Venn diagram* (see Fig 3.4). What is known is that of the 52 cards in the pack 13 are spades and 4 are kings.

Venn diagrams are very useful ways of visually representing a problem so that we can see the relationships between different

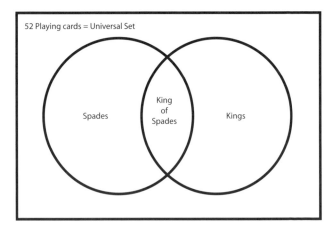

Figure 3.4: Venn diagram of the problem

parts of the problem more easily. So, there are 52 cards in the pack and this is referred to as the *universal set*—i.e. the total number of items. Within the universal set there are two *subsets* the problem is looking at: spades and kings. There are 13 spades in total and 4 kings, but there is an overlap or *intersection* in which one of the kings is a spade.

While 17 cards in the pack will be spades or kings (13 spades: 4 kings), one of these is both. In other words, *the selection of a spade or a king is not a mutually exclusive event*. It is possible to draw a card that is both a spade and a king. The condition of *either* a spade *or* a king would not be met if the king of spades were selected. Remember, the problem was 'spade' **OR** 'king'—not both. So, let's look at what is happening:

1. the probability of drawing a spade is 13/52
2. the probability of drawing a king is 4/52
3. the probability of drawing a king of spades is 1/52—and this would be a failure, so must be subtracted from the possible 'success' of drawing a spade or drawing a king.

The correct calculation for this event is:

$$p_{(S \text{ or } K)} = p_{(S)} + p_{(K)} - p_{(S \& K)} = 13/52 + 4/52 - 1/52 = 16/52 = 0.31$$

This provides a generalised equation for events were the possible outcomes might or might not be mutually exclusive. The general form is:

$$p_{(a \text{ or } b)} = p_{(a)} + p_{(b)} - p_{(a \text{ and } b)}$$

Of course, if **a** and **b** are mutually exclusive, $p_{(a \text{ and } b)}$ is zero and so doesn't count.

The multiplication rule

We may need to consider the probability of the outcome resulting from a chain of events, one following the other. This is often the case in social sciences when we are looking at the possible causal relationship between two factors, such as the impact of sleep deprivation on reaction speeds in tests. An experiment may be set up in which a sample of people are subjected to different sleep patterns, then tested on their reaction times.

If two events are dependent upon each other—that is to say, the outcome of a subsequent event is dependent on the outcome of the previous one—then the probability of those events occurring is calculated in a rather different way. The following simple example illustrates the situation.

Q: What is the probability of spinning a coin twice where a heads is followed by a tails?

Fig 3.5 is the tree diagram for this problem.

The first spin of the coin will produce either a heads or a tails, the probability of each outcome is 0.5. The coin is then spun again and the outcome of that spin is not affected by the outcome of the first: i.e. the coin will still land on either a heads or a tails—the probability of each being 0.5 again. Success relies on the *mutual dependence* of the second outcome on the first for success. If a tail had been

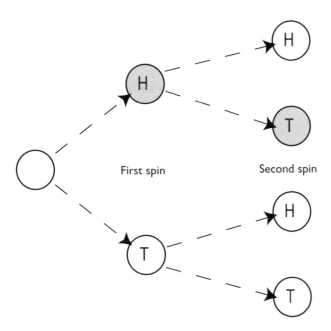

Figure 3.5: The multiplying effect of a chain of events

the outcome of the first spin, the game would be over—i.e. a failure!

However, taking both events together, Fig 3.5 shows there are 4 possible outcomes of the sequence, and only one of these would be a successful outcome—heads followed by tails (the shaded route in the tree diagram). The exploration of a chain of events has a multiplying effect on the number of possible outcomes, with each outcome consequently becoming less likely. Therefore, the probability of success is 0.25 (1 in 4). In these circumstances of *mutual dependence* you determine the probability of the outcome by multiplying the probabilities of each individual part of the event.

$$p_{(H \text{ followed by T})} = p_{(H)} \times p_{(T)} = 0.5 \times 0.5 = 0.25 = 25\% = 1 \text{ in } 4$$

Q: What is the probability of drawing a red card from a pack of playing cards, followed by a black, followed by a red—drawn cards are not put back into the pack once they are drawn?

Suppose you are playing around with a pack of cards and to pass the time you want to know the probability of drawing three cards in the following sequence: red, black, red. After drawing each card it is put to one side away from the pack so that it cannot be drawn again. This now means that there will be fewer cards as you continue to draw from the pack. There are 26 red and 26 black cards to begin with. The probability of drawing a red card first is:

$$p_{(R)} = 26/52 = 0.5$$

There are now 25 red and 26 black cards (51 cards altogether). So, the probability of drawing a black card next is:

$$p_{(B)} = 26/51 = 0.51$$

Finally, there are 25 red and 25 black (50 cards), so the probability of drawing a red next is again:

$$p_{(R2)} = 25/50 = 0.5$$

Notice, in the second draw, there is a slightly higher probability of drawing a black card because there is one more than red. The probability of gaining this sequence is calculated thus:

$$P_{(RBR)} = P_{(R)} \times P_{(B)} \times P_{(R2)} = 0.5 \times 0.51 \times 0.5 = 0.13$$

The odds are stacked in favour of failure. There is 13% chance of success, but there is 87% chance of failure!

Statistical significance

This is a concept we will return to later in the book and it relates to a judgement about whether or not an outcome is likely to be the result of chance. Earlier in the chapter we considered the possibility of throwing a die 100 times with the 6 coming up very time. Seemingly impossible, but each throw of the die is an independent event and so does not—or should not affect the outcome of the next throw. We would calculate the probability of this event as:

$$P_{(6)} = (1/6)^{100}$$

(i.e. 1/6 multiplied together 100 times)

This makes this outcome very unlikely! If it does happen we may not want to believe that this is the result of a 'fix', but we cannot be absolutely certain of this without further tests. In other words, in statistics we are focused on the concept of *uncertainty* and of *statistical significance*. In all research that uses statistical procedures to determine the relationship between events, we are faced with a dilemma.

Q: What is the probability that this event could have happened as a result of chance—i.e. the absence of bias?

A result becomes *statistically significant* if it the probability of it occurring by chance is low. This means that we can begin to suspect an external influence on the result—something is going on—we think! Remember if it is a possible outcome, it may have occurred by chance, so we have to be cautious about stating a relationship between two events when there may not be one.

The concept of *statistical significance* is used to help us judge the importance of what we observe in the social sciences where statistical analysis is the tool for testing possible causal relationships between factors. What we observe may be the result of chance—or it may not. We use *statistical significance* to help us make judgements. We will come back to this many times in later chapters.

Exercises

1. A coin is spun 10 times and results in 9 heads and 1 tail. Should you automatically assume the coin to be biased? What is the explanation for your answer?

2. Are you more likely to draw a black king from a pack of playing cards or a spade of any denomination? How did you determine your answer?

3. What is the probability of drawing a black playing card followed by a red playing card?

4. A student takes a multiple choice question paper as part of a class test. Each question has four possible answers, but only one is correct. There are 10 questions and the student has not done any revision at all. With closed eyes, the student takes a pen and randomly points at an option for each question as the way of answering the questions:

    a. What is the probability of randomly selecting the correct answer for any individual question?

    b. What is the probability of getting every question correct using this method?

    c. Was this a good strategy, or would the student have been better guessing the answers to each question after reading it? Explain your reasoning.

5. Why should you be cautious about stating an event will take place even though the probability of it happening is 99.9%?

6. Probability theory is based on an assumption of 'no bias'. Do you think that such a state of 'no bias' is possible? Support your case with examples.

4 **Probability sampling**

Learning outcomes:

The social sciences are concerned with characteristics of people or groups of people and usually the things that differentiate one group from another. For example, a researcher may wish to compare males with females to see if gender is a significant factor. This will clearly require the researcher to record and measure the characteristic (variable) concerned and then compare the different groups. The statistical techniques for doing this are explained in detail in chapter 9 onwards, which focuses on *inferential statistics* (i.e. statistical techniques we use to explore and explain differences between people). Therefore, the researcher has to be very clear about the design of the research so that the data is collected correctly and in a way that allows valid comparisons to be made. This is the subject of the next chapter. By the end of this chapter you should:

- Understand the concepts of 'population' and 'sample';
- Understand what is meant by 'probability sampling' and why it is used in the social sciences;
- Be able to define 'sampling frame' and identify the problems associated with generating samples;
- Understand the principles underlying: simple random, stratified random, and, multi-stage cluster sampling;
- Understand the importance of sample size and the likely impact on accuracy of research findings;
- Understand the difference between certain types of non-random sampling (e.g. opportunity sampling) that appear to be random, and genuine random sampling.

Populations and samples

Consider the following simple research questions:

> Q1: Is there a difference between the initial reaction times of male and female drivers in the UK when reacting to a driving emergency?
>
> Q2: Are the average earnings of 21 year-olds in full-time employment linked to the number of years they have been working and their qualifications?

Both questions require the researcher to identify:

A. The variables to be measured;
B. The group of people who will be the subjects of the research.

Chapter 5 will look closely at 'A' and how you arrive at the variables to be measured, as well as how to establish the general ground rules for designing a research project.

Population

The group of people who are the subjects of the research is called the *population*. In question 1, **all** drivers are included (as all are either male or female). In question 2 **all** twenty-one year old full time workers. However, the researcher is presented with a number of issues to resolve because defining the population is not always quite so easy. For example:

- In question 1, what is meant by 'drivers'? Is this anyone with a driving licence even if they are no longer driving regularly, or possibly no longer active drivers?
- In question 2, what do we mean by 'full-time employment'?

When we identify and describe a population it is crucial that two conditions are met:

1. There must be clear criteria that define the characteristics of the population so that there is no ambiguity about who is or is not included;
2. Everyone who meets the criteria is included in the population and everyone else is excluded.

In general everyday usage, we tend to refer to *populations* in the sense that they are all the people who live in a particular area or country. In the social sciences, we are being more specific and technical about the use of the word to describe the group we are studying. The population may be relatively small, such as all the students registered at a particular university or large as would be the case with the two questions posed earlier. Also, the research population may be a sub-group of a much larger population (which we call the *universal population*). For example, Fig 4.1 illustrates the population linked to question 1.

In Fig 4.1, the research population is represented by the white oval area. It is a clearly defined group of people drawn from within the general UK population—i.e. they are a subset of it. Everyone who meets the criteria set out in the oval area is included in the research population. The shaded area represents the rest of the UK population that does not meet the criteria and so will be excluded from the study.

Being unclear about the research population is a common weakness with inexperienced researchers and is also a common cause for criticism of work. This is sometimes made worse when the research results are used to

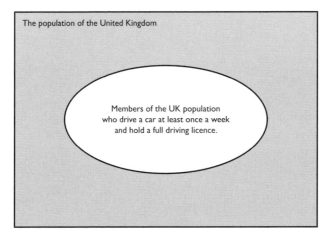

The population of the United Kingdom

Members of the UK population who drive a car at least once a week and hold a full driving licence.

Figure 4.1: Identifying the research population

draw conclusions about specific groups of people. If the research population is not clearly defined and selected, the work falls at the first hurdle.

Sample

In an ideal world, every member of the research population will be included in the research. If there are 25 000 000 drivers then the ideal situation would be to design an experiment to measure reaction times and test all 25 000 000! It should be immediately apparent that this is not a feasible proposition. The resource implications of such a task are quite extraordinary. There has to be a compromise somewhere that lets you research the problem but to do so within the resources at your disposal.

The approach is to survey a much smaller group of people selected from the population—i.e. a *sample*. Fig 4.2 illustrates the idea of a sample drawn from the population of drivers Identified from Q1.

Opinion polls are a typical example of this problem. Consider the following question that may be asked:

Q3: If a general election was called for today, which political party would you vote for?

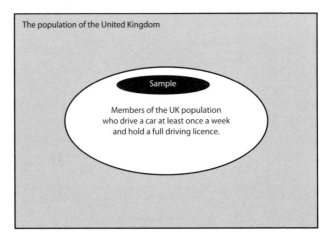

The population of the United Kingdom

Sample

Members of the UK population who drive a car at least once a week and hold a full driving licence.

Figure 4.2: Representation of a sample drawn from the research population

The research population for this question will be all adults registered to vote. Again this will run into millions and the only way for pollsters to gather this information is by asking this question of a *representative sample* of the electorate.

Consider the following question:

> Q4: Is the memory of young children improved by the addition of Omega 3 rich oils in their diet?

Even if 'young children' is clearly defined during the research design stage, there are a lot of young children in the general population! However, creating a clear definition of which young children the researcher is interested in (e.g. primary school age) does create an obvious population. This will also be a very large number when considered at a national level, so the researcher will need to create a *representative sample* from the research.

A sample is generally a considerably smaller number of individuals than the research population itself, yet researchers use results gained from the sample to make claims about the research population as a whole. Naturally, this presents potential difficulties. If the sample is a small group selected from the research population, how can the researcher be sure that the findings gained from this group can be extrapolated to the whole population?

The short answer is that there can be no certainty about the ability to generalise results from a sample to the population as a whole.

The principle behind drawing a sample is that it is supposed to be *representative* of the population as a whole and, in theory, conclusions about the sample can be extrapolated to the population as a whole. The technique used to generate representative samples is called *probability sampling*.

As you might expect, the larger the sample (i.e. the greater the proportion of the population selected to be in the sample) the more likely the results will reflect the reality of the population—but more of this later in the chapter.

Probability sampling

The point of carrying out social and psychological research on a *sample* drawn from the population is that the findings gained from the sample apply to the population as a whole. This is perhaps a tall order since samples often involve a relatively small number of people relative to the population. Logic would suggest that the composition of the sample must reflect the characteristics of the population from which it is drawn—e.g. age range, gender balance and any number of other characteristics considered necessary to make the sample a *representation* of the population.

What is probability sampling?

In chapter 3 we considered the theory underpinning probability. Probability sampling uses this theory to generate a sample that we hope will reflect the characteristics of the population. If we remove all sources of *bias* from the selection process then each person in the population has the same chance of being selected as any other. Going back to the playing card scenario in chapter 3, the ace of spades has a 1 in 52 chance of being selected if picked *randomly* from the pack (i.e. the person selecting the card has no idea which card is being picked as all cards are placed faced down, for example). The probability of a red card being selected is 1 in 2.

The principle of probability sampling is quite straightforward, if the sample is generated by random selection, the characteristics of the sample group are likely to be in the same proportion as those found in the population as a whole—in other words, the sample is *representative* of the characteristics we are focusing on. Fig 4.3 illustrates this as it applies to Q3 above (i.e. reflects the voting intentions of the population).

This is clearly a simplistic example, but if probability sampling has worked, then in theory the sample should reflect the voting pattern of the population of registered voters—and party A will receive the majority of votes—i.e. the sample is *representative* of the population of registered voters.

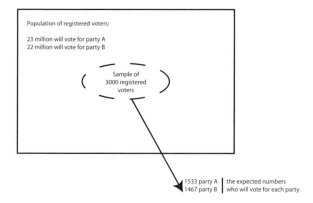

Figure 4.3: Theoretical sample produced by random sampling

Problems associated with probability sampling

However, as we saw in chapter 3, because each individual has an equal chance of being selected it is perfectly possible (if unlikely) that the sample could produce 3000 people who will vote for party B, or 2000 who will vote for party B, and 1000 for party A. If that were to happen (and it does in the real world) we get a false result due to *sampling error*. We will return to sampling error several times throughout the book and it is something researchers always have to keep in mind and attempt to compensate for when reporting their results.

In other words, probability sampling **may** result in a sample that is not representative of the research population simply because of the random nature of the selection process, and this is a paradox we need to avoid. Look at the possible following scenarios:

A. In a research project to identify if gender influences the ability to multi-task, a sample of 1000 people is generated—all of whom turn out to be male!

B. In a research project to identify if age influences voting intentions, a sample of 1000 people is generated—all of whom are aged between 25 and 35!

In scenario A, if there is only one gender in the sample, the research is invalidated—there will be no data on female ability to multitask. In scenario B there is only a 10 year spread in the sample—there will be no data on voting intentions of younger or older voters who collectively span a much greater age range.

There are a number of methods used to attempt to reduce *sampling error* and their effects.

Techniques of probability sampling

There are three principal techniques employed in probability sampling and they are explained in detail here, but first it is essential that the population is clearly defined and those meeting the criteria have the possibility of being selected to be part of the research sample. This is done by creating a *sampling frame*.

The sampling frame

The short explanation of a *sampling frame* is that it is a list of **every** member of the population that meets the selection criteria. This sounds simple enough, but consider the following question:

Q5: Is brain development in young children adversely affected by close proximity of their homes to urban motorways and other major roads?

The point here is that defining the population to be investigated cannot be separated from the research design process—and the link to chapter 5 is very strong here. The *sampling frame* will be largely determined by the way the research is designed. In the case of Q5 it is not just a matter of identifying young children living in properties located close by (and what does that mean?) an urban highway, you would also need to identify young children living in properties that are not located near an urban highway (for comparison purposes). In other words, there are both design and practical difficulties to consider.

Once a population has been identified, its members are listed and given a unique code that identifies them. Fig 4.4 is a simple example of an extract of a school's population. In this case, every student has a unique *identifier*—the enrolment or

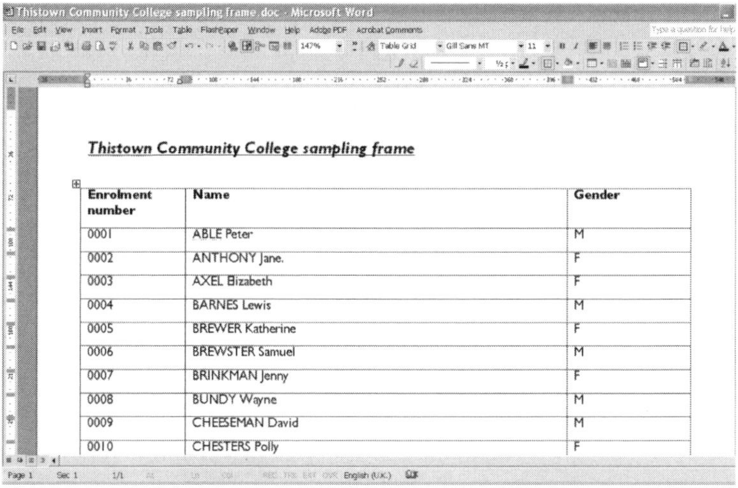

Thistown Community College sampling frame

Enrolment number	Name	Gender
0001	ABLE Peter	M
0002	ANTHONY Jane.	F
0003	AXEL Elizabeth	F
0004	BARNES Lewis	M
0005	BREWER Katherine	F
0006	BREWSTER Samuel	M
0007	BRINKMAN Jenny	F
0008	BUNDY Wayne	M
0009	CHEESEMAN David	M
0010	CHESTERS Polly	F

Figure 4.4: Creating a sampling frame

registration number. In the absence of any obvious identifier, the researcher would simply create one, the simplest being by sequential numbering. In some cases, the population may be all the individuals recorded in a database (such as a school).

From Fig 4.4 it is clear that a sampling frame is simply a listing of the research population where each individual is given a unique identifier. Note also that gender is recorded in this case. The sampling frame may also contain other significant information that may be of relevance. It may be that the topic being researched here is something linked to gender differences. We will come back to this in *stratified random sampling*.

Having identified the population and created the sampling frame, we can now begin the sampling process. There are three main ways of sampling, each suited to a different purpose:

- *Simple random sampling*: The most straightforward way of sampling, this is used where there are no requirements to ensure a range of factors such as gender balance or ethnicity need to be accounted for.

- *Stratified random sampling*: If we need to ensure a particular characteristic such as gender, age or ethnicity needs to be properly represented in the sample, the sampling frame is sorted and subdivided before sampling from each sub-set.
- *Multi-stage cluster sampling*: This is frequently used on large-scale social sciences research projects where complex combinations of characteristics need to be adequately represented in the sample.

Simple random sampling

Once the sampling frame has been constructed the sample is drawn by random selection from that list. So, for example, if the sampling frame has 1000 entries and you want to create a 10% sample (and we will look at the significance of sample size towards the end of the chapter), 100 individuals need to be selected. Theoretically, since every entry in the sampling frame has an equal chance of being selected, the final sample is, we hope, representative of the population.

However, you still need to apply a system that ensures randomness, and which minimizes bias. Simply listing all the names and then selecting every tenth name may look *random*, but who wrote down the list and was this in alphabetical order, or organized in some other way—deliberately or otherwise? The accepted way of selection is by assigning a number to every individual subject of the population. A table of random numbers is then used to select your sample.

Fig 4.5 is part of a table of random numbers. You can buy tables but it is very easy to create a table in Microsoft Excel. Fig 4.5 was generated in Excel.

Using the random numbers table

The next stage is to use the table of random numbers to identify the subjects for the sample. There is no science involved here. Close your eyes and stick a pin on the page! Whichever number it hits is your starting point. You now work horizontally or vertically and record each number in the row or column. For example, in Fig 4.5 665 has been

Random Numbers 1 to 1000

105	608	866	460	833	999	245	546	160	176	939	794	238	802
718	243	252	141	767	869	370	317	681	979	146	43	628	777
950	325	769	505	402	73	513	202	741	486	410	738	841	861
429	762	900	867	894	874	782	31	943	248	22	716	332	205
117	408	166	460	875	452	633	469	973	632	35	885	358	688
385	115	452	246	416	451	503	244	727	331	429	602	747	349
476	175	550	676	46	738	624	433	127	980	860	240	720	146
124	542	71	672	162	165	88	509	23	26	968	628	451	377
117	3	843	618	374	389	256	254	165	7	574	122	24	212
717	838	357	363	568	533	961	560	782	740	634	300	878	265
875	443	282	881	938	540	162	271	14	399	710	315	297	859
766	233	253	37	349	971	135	769	774	386	176	504	667	904
555	606	1	232	111	753	640	748	891	50	305	810	793	703
841	261	835	912	840	326	722	405	345	515	728	643	864	654
791	949	37	998	845	838	68	273	77	690	584	992	956	854
521	827	740	812	530	495	205	846	573	842	883	109	909	213
815	92	3	265	660	718	711	445	908	983	710	855	200	900
947	212	76	368	142	451	79	614	594	623	218	551	455	157
240	242	826	755	237	292	186	581	134	225	322	634	594	647
217	990	421	263	404	418	682	848	89	537	392	12	52	996
791	633	41	592	708	973	355	730	395	627	198	199	969	126
689	154	99	570	453	767	586	549	513	814	667	956	295	542
692	710	846	112	665	972	324	723	131	593	450	665	936	491
297	309	889	627	166	547	823	759	63	884	911	353	748	724
853	102	858	536	787	8	389	246	992	717	193	174	760	655
819	227	161	115	28	340	559	402	798	558	425	65	898	565
390	764	445	718	665	461	397	38	687	716	696	830	897	392
141	238	129	779	745	822	458	59	520	109	417	437	490	966
821	843	588	900	122	259	879	455	879	353	630	301	42	246
546	578	881	185	651	53	938	750	97	378	16	753	989	732
221	705	121	351	190	740	959	641	603	649	181	139	743	645
305	665	392	711	24	160	8	192	730	891	363	57	972	

Figure 4.5: Random numbers tabled generated in Excel

circled and this identifies the entry in the sampling frame with that number. Working downwards, the next number is 166, and so on until you reach the bottom of the column. Where a number is repeated, it should be skipped and you move on to the next. When the bottom of the table is reached, another starting point is selected as before. In this example, 560 is circled as the second starting point, and the direction is taken as horizontal. This process is continued until the sample size has been reached.

Stratified random sampling

The problem with simple random sampling lies in one of the basic difficulties of probability sampling in ensuring the sample is representative of the population from which it is drawn with respect to the characteristics being explored. Consider the following question:

Q6: Are female history students better at memorizing historical dates than male history students?

To answer this question the researcher will need to compare the performance of females with males. The risk with simple random sampling is that the sample drawn may not have a sufficient balance of males and females. To reduce the risk of this you would *stratify* the sampling frame. Looking at Q6, it is important that the sample is representative of the gender balance in the research population—because this is a key variable to be researched. For example, if your population has 30 000 subjects, of which 10 000 are female, the sample needs to be representative of this gender balance. If a 10% sample is to be taken—i.e. 3000—you will need 1000 female and 2000 male history students.

You split your population into the strata you are trying to properly represent—in this case, gender. In effect you generate separate sampling frames for each strata (gender). The simple random sampling method is now applied to each stratum so that 1000 female and 2000 male history students are selected.

It is possible to use stratified probability sampling using more than one factor. For example, you may want to know if attitudes to professionalism are related to gender and/or educational level of entrants. Figure 4.6 illustrates how the population has been stratified into four strata.

female graduates	**male graduates**
female non-graduates	**male non-graduates**

Figure 4.6: Stratifying for multiple criteria

Multi-stage cluster sampling

At first sight this may look like stratified sampling, but the similarity is superficial. A typical scenario for this form of sampling might be a study of the psychological impact of the fear of crime in the UK. In considering your sample, a number of factors have to be be taken into account to gain a good representation of the population. There is a wide geographical spread, a range of urban and rural settings, a range of economic and social conditions at the ward level within a local authority. You can't easily create a sampling frame of the whole population.

Representation can be achieved by *multistage cluster sampling*. Fig 4.7 represents the approach.

The ultimate aim of sampling is to create a representative sample of households across the country, taking account of location and social composition. This is a very complex and difficult thing to do, but *multi-stage cluster sampling* addresses the problem. In the example shown in fig 4.7, stratified random sampling is used to create a sample of urban and rural local authorities. Within each of these samples, stratified sampling is used to generate a sample of wards in each category of urban and rural. Samples of streets or neighbourhoods representing a range of social indicators are generated from each ward selected. Finally, individual households are identified and sampled through simple random sampling.

This form of sampling demands a very high level of organization and is usually employed with very large national research projects.

Sample size

Now we come to the vexed question of how large a sample should be. There is an answer based in theory, and another based in practicalities. Let's take practicalities first.

Research costs money and requires time and physical resources (e.g. paper, postage, telephones and so on). Often, the constraints set by the available funding and other resources determine the sample size. However, as a general

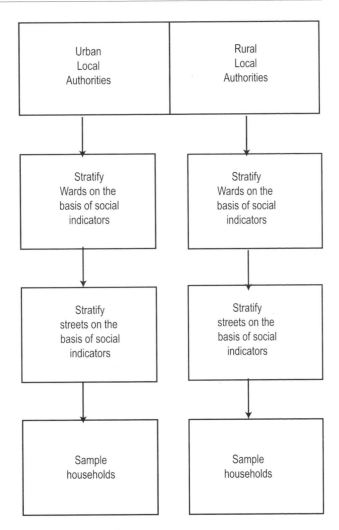

Figure 4.7: Multi-stage cluster sampling

principle you should always go for the largest sample possible. Consider the logic of this. Remember, the researcher is attempting to draw conclusions about a population and the sample is meant to represent the population and the statistics of the sample are used to describe the population in general. However, the sample is not the population and so you cannot be absolutely certain that the sample statistics (e.g. average salary) apply to the population because of the potential for sampling error. A simple case will illustrate the issue of sample size.

> Scenario: There are 5 people and their salaries are £20,000, £25,000, £22,000, £30,000 and £35,000.

The average (mean) salary of the population (i.e. of the 5 employees) is £26,400[1]. However, suppose the average salary of the population is to be estimated from a sample of 2 employees. If random sampling results in selecting the employees earning £20,000 and £35,000, the average salary of the sample is £27,500. If we now assume this represents the average salary of the population, the researcher would be making an error. The larger the size of the sample relative to the population, the smaller the sampling error is likely to be. We will be looking at the reasons for this in chapter 6.

This brings us neatly to the calculation of sample size from the theoretical perspective. It is problematic because you need to work back from the error you are prepared to accept! The larger the sample the more confidence you can have in your findings. There is a formula for calculating the theoretical minimum sample size and this is based on a couple of concepts that will be introduced in the next chapter, so don't worry at this stage if their meaning is not clear here. The sample size is calculated using the formula:

$$n = \frac{s^2}{(S.E.)^2}$$

n = the sample size required
s = standard deviation of the variable being investigated

---

[1] We will look at the mean and other *measures of central measures* in chapter 6.

S.E = standard error of measurement (a measure of the degree of accuracy in the result considered acceptable by the researchers)

NOTE: The concepts of standard deviation and standard error of measurement will be explained in chapter 6.

Non-probability sampling

There are several sampling techniques that appear at first sight to be random sampling, but they are not yet they are often used because of heir convenience.

If you have ever been stopped in the street by a market research interviewer, you have been *opportunity sampled*. Interviewers are briefed to identify passers-by who conform to a given set of criteria. The interviewer stands in a particular spot and selects us because we conform to the criteria. This is not random sampling: we have consciously been selected by the interviewer, and we consciously decided to pass that way (although not with any expectation of being selected). Inclusion in the sample is therefore not the result of a random event, but the result of *opportunity sampling*. It is used because it is convenient.

Another technique used erroneously by inexperienced researchers is to use a list that is already available (e.g. a school register or electoral roll), which is fine, but it is how the list is used to generate the sample that may be a problem. If the list has 1000 names and 100 are needed in the sample, the researcher simply selects every tenth case on the list. Again this is not random because selection is pre-determined by the position in the list. If the list is alphabetical, then this provides an additional pre-determining factor.

Exercises

1. Explain how you might produce a sampling frame for the the following research projects:
   a. Do female learner drivers make fewer errors than men in their driving test?
   b. Do female university students tend to achieve better degrees than male university students?
   c. Do professional athletes have better spatial aptitude than the general population?
2. Suggest the most appropriate method of probability sampling for each of the research projects in question 1. Explain the reasoning behind your decision?

5 Defining and refining your research

Learning outcomes:

Statistical analysis is not just a simple question of number-crunching. The numbers have to be collected using methods and approaches that give a high level of confidence in what the numbers tell us. This chapter looks at some of the key areas of research design that matter when collecting primary data (i.e. information directly from observation—also called empirically derived data). By the end of the chapter you should:

- Understand how to develop research questions into statistical variables for data collection;
- Understand the concepts of 'reliability' and 'validity' and their importance in designing social research activity;
- Understand the basic principles underpinning questionnaire design.

Hypothesis to statistical variable

Social research is underpinned by the ideas that arise from *hypotheses*. Look at a simple example to show how it will lead to research activity. A researcher may make the following statement:

Migrant workers are usually employed in lower skilled occupations than those for which they are qualified.

A hypothesis may be a belief, or an assertion that may or may not be based on any observable facts. In other words, it can come from anywhere! The important feature of the hypothesis is that it can be tested through research—a process called *hypothesis testing*. Other hypotheses are the direct result of research activity, and may be a by-product of research into something else—the researcher noticing some

Figure 5.1: The research journey

unexpected information and so develops a hypothesis to be tested later—a process called *hypothesis building.*

In most cases, research is designed to test hypotheses, and there is a journey to be made from hypothesis to the outcome of research data analysis. Fig 5.1 illustrates the step by step process.

In this chapter, the process will be considered insofar as it leads to the collection of quantitative data for statistical analysis (qualitative research will follow a similar path but clearly the data collected will not be based on numbers and will be analyzed in a very different way). You should also note that the top of the ladder does not use the work 'proof'. Your data will both support your hypothesis, and answer the research question, or they will not. They cannot prove it because there may be factors that you have not been able to control, or of which you have no knowledge. More will be said about this under Research Design.

Research question to research variable

The hypothesis is a statement, an observation or assertion that will need to be tested through research. However, the hypothesis itself needs to be expressed in a form that can be used to design the research—the *research question.* Research is meant to answer, at least in part, some question that we are posing. The question helps the researcher to:

- Define the limits of the research—i.e. its scope
- Identify and define the *concepts* that must be explored in the research and which lead to
- The variables the research will collect data on—i.e. measure.

The research question

The problem with the hypothesis that began this section of the chapter is its breadth. It is difficult to design the research directly from the hypothesis as this is not a question. Also, the general nature of the hypothesis makes it difficult, so we may wish to refine this into a more manageable research-focused question such as:

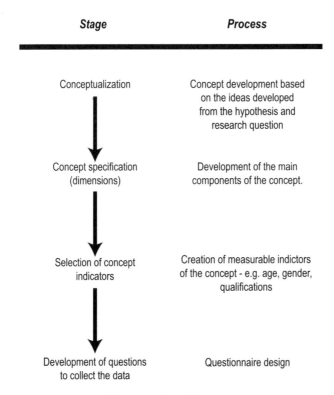

Stage	*Process*
Conceptualization	Concept development based on the ideas developed from the hypothesis and research question
Concept specification (dimensions)	Development of the main components of the concept.
Selection of concept indicators	Creation of measurable indictors of the concept - e.g. age, gender, qualifications
Development of questions to collect the data	Questionnaire design

Figure 5.2: Operationalizing a research question

> Research question: Are *migrant workers* in *Exeter* from the *EU Accession States employed* in occupations with lower skill levels than those for which they are *qualified* in their *home countries*?

This question is now very specific and can be answered if the research is properly conducted. The highlighted text in the question are the key *concepts* that we need to define and from there you can develop the specific questions used in the research to collect the data. The process of converting the research question into measurable variables is called *operationalization* and Fig 5.2 illustrates the processes involved.

Conceptualizing

The emboldened text in the research question is a list of *concepts* (i.e. ideas or observations that relate to the way we perceive the world around us) that need to be clarified so that there is a clear understanding of what they are. Consider these in turn.

- *Migrant workers*: We need to define these so that they can be differentiated from the working population as a whole. In this case, we are defining them as workers of another nationality who are now living and working in this country.

- *Exeter*: This may seem an obvious concept but there may be different interpretations of the geographical limits. Do we mean the City of Exeter, or the wider travel to work area? Also, this has a bearing on our definition of who to include in our sample of migrant workers. Are we concerned with migrant workers who are currently resident in Exeter, or who may live outside of Exeter but travel to Exeter to work?

- *EU Accession States*: This is clearly defined as those who countries that joined the EU in 2004. This is an easy concept to define and it also highlights the importance of such concepts as the results of this study can be compared with similar studies elsewhere on migrant workers from the Accession States (often referred to as the A8 states).

Where there are formally agreed definitions that are reflected in the broader research literature, they should be your default definitions, unless they present some issue linked to what is being researched. For example, the research hypothesis may be that these definitions are themselves flawed or inadequate in some way.

- *Employed*: Are we including unpaid, full- and part-time work? Does this include self-employment?
- *Qualified*: Do we mean educational level as compared with the equivalent UK qualifications, or do we mean qualifications that entitle them to practice a trade or profession (i.e. vocational qualifications), or do we mean both? We would need to consider this very carefully. For example, being a graduate does not automatically imply being qualified to do particular jobs, and that applies to UK nationals.
- *home counties*: There are clear definitions used by national government agencies in reporting statistics relating to migration and social policy. We are likely to define this as the individual's nationality of birth or naturalization. When combined with the definitions of migrant workers and EU Accession States, the three concepts allow the researcher to clearly define the potential research population.

Concept specification and indicators

It is easier to explain this by example. Consider the concept of *migrant worker*. The factors or *dimensions* of the concept will include:

- Age
- Nationality
- Current residency
- Home residency
- Work.

The *indicators* that will then define this group and then measured by collecting the data via the research activity, which may be a questionnaire completed by the person or by the researcher during an interview. The next chapter looks at questionnaire design. They also form the *variables* that will be used in the analysis alongside the other variables for the other concepts identified in the research question. The other measurable variables in this research may be:

- Current occupation.
- Qualifications—although this may be divided into several variables to capture the range of academic and vocational qualifications that may be needed to answer the research question.
- Occupation in their home country—this may help to relate unfamiliar qualifications to employment in their home country.
- Current residence location (e.g. by postcode) or that of their employment (if employment location is the key indicator rather than were the person lives).
- Employment status in this country.

Validity and reliability

These two concepts are one of the chief weapons of criticism of a social research project by its detractors. Get these wrong when designing your research, and there will be no hope of acceptance of your findings and analyses.

Reliability

The data collected, and by implication the methods used to collect it, are *reliable* if the responses to questions are consistent. For example, if a questionnaire is given to two randomly selected samples of individuals from the research population, we would expect the responses to be comparable—after all both samples are supposed to be representative of the population. If they are, the questionnaire is generating *reliable* data. *Reliability* is a measure of its repeatability or stability. In psychology, for example, psychometric instruments are tested in this way to make sure they are reliable. This particular test of reliability is called **test-retest reliability**.

Another test of reliability is whether or not respondents (the name given to those answering a questionnaire) interpret the questions in the same way. For example, consider the following question:

Q: Where do you live?

The problem with the question is that people will interpret it in different ways. Just consider these possible responses:

- I live in a house
- I live in England
- I live in the UK
- I live in Thistown
- I live in This Street .... and so on!

All of these responses are *valid* (see below) in that they do represent the real places where a respondent lives. The difficulty is that it does not result in a consistent interpretation of the meaning of the question—the wording is too open to individual interpretation. The question is therefore unreliable. Now look at the alternative wording of this question:

Q: What is the first part of the postcode of your address (e.g. AB12)?

The meaning behind this question is now clear to everyone reading it and there is even an example to make sure! This will result in an increased reliability of response to the question. There is more on questionnaire design later in the chapter, but for now it is important to grasp the key point that the reliability of data frequently hinges on the quality of the question.

It is common practice to *pilot* a questionnaire using a small sample of individuals to check its reliability and identify any other issues (e.g. the language used may not be at an appropriate level). This allows the researcher to modify the questionnaire before applying a flawed version to the research sample. It is also general practice that the members of the pilot sample are excluded from the final research sample. This is not usually a problem as the numbers in the pilot are generally small in relation to the research population.

Validity

Validity is the ability of your research to gather information on the variables it claims to be measuring. This is a complex issue and is dependent on a number of factors. For example, if you have operational zed the concept of **employment** for the hypothetical study of migrant workers in Exeter described earlier by identifying its dimensions, you will then construct questions to collect data. The validity of your questionnaire will depend on:

- An accurate operationalization of the concept—i.e. what do we mean by **employment** and is there a universally accepted definition?
- The structure of the questions and the questionnaire are such that respondents understand what you are asking, and there are no ambiguities in wording or structure—i.e. here is the link with reliability since unreliable responses lead to invalid data interpretation and reporting.

It is possible to design a questionnaire that is *reliable* because the responses are consistent, but it may be *invalid* because it is not measuring the concept you think it is measuring. *Reliability* is not a measure of *validity* but it can influence the *validity* of your findings. It is beyond the scope of this book to go into detailed discussion on how to test for *validity*, but here are a couple of practical tips to help you at least carry out a simple assessment.

- If you have developed a scale, rating or ranking question (see later in the chapter), discuss its structure with others who know about the particular concept to see if they recognize and agree with your analysis.
- Test the question on a few people who you would expect to score highly, and on a few from whom you would expect a low score and see if their responses are reliable enough to suggest your questions are valid.

This is not a guarantee, but obvious errors should be identified. To test a questionnaire for reliability and validity is both time-consuming and expensive and usually involves large-scale piloting and comparisons with other questionnaires known to be *reliable* and *valid*.

However, what this description does tell you is that when it comes to looking at other researchers' data, you also need to ensure that their methods were *valid* and *reliable*, and it is a good idea to read the research papers of experienced researchers to see how they argue their case for validity and reliability, and of their critics who may argue that they are not!

Questionnaire design

Questionnaires are the principal method of collecting primary data in social research. This may be administered to members of the research sample (e.g. via postal survey) or used to record data during one-to-one interviews (face to face or via telephone survey).

Structuring the questions

The type of response required of each question can affect the way a respondent answers it. If a respondent cannot understand the question or finds it difficult to take it seriously, there may be no response or you will find sarcastic or other derogatory comments written in. With the best will in the world, you will always find a minority of respondents who add comments to your questionnaire because they did not like a question or preferred to provide a response of their own! We want this to remain a tiny minority, so great care should be taken in designing a questionnaire that will be taken seriously because the questions are well structured, logically sequenced and do not form part of a questionnaire that is over-long and tedious!

The other point to be borne in mind as a researcher is that the data will be entered into a database such as SPSS for detailed analysis, so it is important that the layout of the questionnaire facilitates rapid and accurate transfer. Commercially, this may be achieved by creating questionnaires that collect data via an Internet based form—with direct entry to the database as a respondent works through the questions—or via forms that may be electronically scanned into the database. While the cost and accessibility of this technology is reaching the point where this is a feasible proposition for the lone, freelance researcher, it is not quite there yet. In any case, it does require some technical knowledge on the part of the researcher and for students, the old fashioned method of manual data transfer from the questionnaire to the database continues to be the norm.

Open or closed responses?

You are going to require one of two forms of response—*open* or *closed*. A *closed* response question provides data that are easily coded for entering into a database, whereas an *open* response can often be very difficult to code. The examples given below illustrate the difference between the two.

Closed response

A typical example of this type of question might be:

Q: From the list below, tick the <u>one</u> item that best describes the most important reason for enjoying your work	✓	
I enjoy the variety of work		(1)
The working hours suit my circumstances		(2)
The salary is good		(3)
Other reason (please state reason below)		(99)

Each option will be given a code and this makes the job of data entry much easier. Note the 'Other reason' option. This can be a useful device to create new coded categories that you had not originally thought of. This is coded 99 to allow you to create new codes from 4 to 98 (if needed). You would do this by looking at all the responses to this question to see if there are any reasons that are repeated sufficiently often to justify a separate code. This relates to the *coding frame* approach discussed in chapter 2. This makes it very easy to transfer the code to the database. Fig 5.3 shows how this question relates to the variable **enjoy** in a hypothetical study of attitudes to work in SPSS.

Figure 5.3: The relationship between a question and the variable it represents

Note that the researcher has identified a fourth item where quite a number of respondents have reported the social aspects as important in the 'Other' category. This has occurred sufficiently frequently to justify a new category as it is clearly important to many people. This illustrates the value of allowing respondents to describe their 'Other' option and giving leeway to generating new category codes.

Open response

This approach is useful if you do not wish to impose categories of response, or if you are on a 'fishing trip' in an attempt to identify possible categories where you have no clear idea of what they should be. Using the question in the last section, the following might be the *open response* version of it.

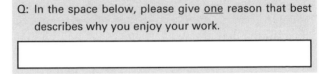

Q: In the space below, please give <u>one</u> reason that best describes why you enjoy your work.

This allows respondents much greater control over what they can report. The advantage of this approach is that the information will be richer and more informative than imposed categories that may not quite match up to respondents' actual experience. The disadvantage is that it could present you with a coding nightmare. Each response will need to be looked at and a category created. An alternative, and more practical approach, is to look at common areas and create coded categories that encompass all the responses. Whatever approach you take, *open response* questions, while richer in terms of the diversity of information gained, do take a great deal of work to analyze later.

Single or multiple response?

The two previous examples are called *single response* questions because the respondent can only choose one item. The question types discussed in chapter 7 are good examples of *multiple response* questions where respondents are asked to provide more than one response to the question asked. The use of scales, ratings and rankings require the respondent to respond to every item, but there are other ways of using a *multiple response* question. For example:

Q: From the list below choose <u>up to three</u> reasons that most influenced your choice of university course. Place them in order of importance from 1 to 3. (Please note, <u>three</u> is the maximum number, but you can select one or two if that accurately reflects your choice).

This question has the effect of letting the respondent know that you are acknowledging there may be more than

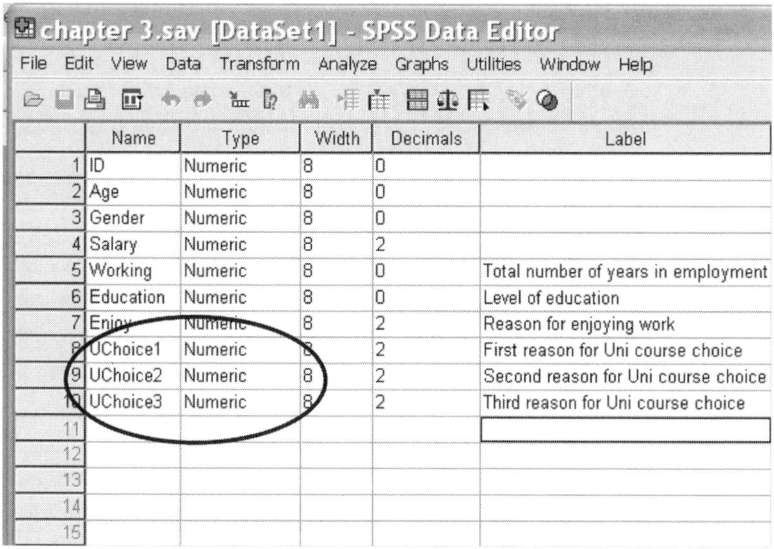

chapter 3.sav [DataSet1] - SPSS Data Editor

File Edit View Data Transform Analyze Graphs Utilities Window Help

	Name	Type	Width	Decimals	Label
1	ID	Numeric	8	0	
2	Age	Numeric	8	0	
3	Gender	Numeric	8	0	
4	Salary	Numeric	8	2	
5	Working	Numeric	8	0	Total number of years in employment
6	Education	Numeric	8	0	Level of education
7	Enjoy	Numeric	8	2	Reason for enjoying work
8	UChoice1	Numeric	8	2	First reason for Uni course choice
9	UChoice2	Numeric	8	2	Second reason for Uni course choice
10	UChoice3	Numeric	8	2	Third reason for Uni course choice
11					
12					
13					
14					
15					

Figure 5.4: Creating variables in SPSS for a multi-response question

one reason influencing a choice. By restricting the choice to three, you prevent the respondent from selecting everything, which would make the question useless. By forcing a limit, the respondent will, you hope, identify the most important reasons. This question will generate three variables in the database, e.g. **UChoice1, UChoice2, UChoice3**, reflecting the order of importance. Fig 5.4 shows how this may look in the SPSS database.

Structuring the questionnaire

The overall look of a questionnaire and the ordering of the questions are very important. You need to maintain respondent motivation, especially with a long questionnaire. Make it easy for a respondent to want to participate by using progressively graded questions that lead the respondent through the process.

The questionnaire structure can have a large influence its reliability and validity. You always have to keep in mind that the person reading and responding to your questions is interpreting them in their own terms. The order of questions can send out unintentional messages about perceived intentions. In particular, the response to particular questions can be affected by the nature of the preceding ones. There are several techniques used to mitigate their effects.

Funnel technique

The questionnaire begins with broadly based questions and gradually works towards questions that have a narrow area of focus and is at the core of what you are researching. The *funnel* technique is used:

- where respondents are generally well-motivated;
- where the purpose of the survey is to obtain detailed information (the funnel approach helps respondents recall detailed information more efficiently);
- where a particular question may impose a frame of reference before gaining respondents' views and so unduly influence the responses.

The following is an example of a simple *funnel* sequence of questions. Please note that they are not written in the way they would be presented in a questionnaire and serve only to illustrate the approach.

1. List some of the most important problems facing the European Union?
2. Which one of these problems do you think is the most important?
3. Why do you believe it is the most important?
4. Where do you find most of your information about this problem?
5. Which newspaper(s) do you read?

The overt agenda of this sequence is to see how opinion is related, if at all, to the sources of information and, more particularly, the newspapers read by respondents. The funnel sequence helps the respondent to focus on thinking about the problems, as a sort of warm up to asking the target question of what they believe is the most important issue, and how they are informed about this issue. The newspaper question comes last because if it came earlier, it may bias the response to questions 3 and 4 as the respondent may make all sorts of assumptions about the questionnaire's agenda.

Inverted funnel

As the name implies, the *inverted funnel* approach reverses the sequence. It starts with the narrower questions and develops into a broader view. This sequence is used:

- where the subject matter is not likely to be very motivating, or is of no particular importance or interest to respondents (i.e. a respondent is more likely to answer questions that are specific and easy, rather than generalized ones that may require more effort);
- where respondents' experiences of the subject matter is not very recent and memory may be hazy and unreliable (specific questions will help a respondent recall facts etc);
- where the respondent will be asked to make an overall judgement about facts (i.e. narrow questions can establish facts first of all).

The following is an example of *inverted funnel* sequencing. Again, the questions are not meant to be taken as perfect examples, but to illustrate the process.

1. Which of the following issues did you discuss with your adviser?
2. After your discussions with your adviser, which of the following actions were you able to take?
3. Did these actions in general have the desired outcome for you?
4. How would you rate the quality of the service you received from your adviser?

In this example, the target question is a general rating of service quality. However, as this type of question is frequently asked some time after the event concerned, it is often necessary to make the respondent recall the specifics of the event first so that a considered judgement is more likely to be made.

Filter questions

Some questions or sections of your questionnaire may apply to some respondents but not others. Under these circumstances it is necessary to divert respondents away from inappropriate questions towards other parts of your questionnaire. You use a filter question that identifies which respondents need to answer which questions. The following is an example of this.

1. Do you live alone? If NO go straight to question 2. If YES go straight to question 4.
2. How many other people live in your household?
3. List the number of people in each of the following age groups?
4. On average, how often do you visit your local public library?

Question 1 is the filter question that directs respondents to the appropriate parts of the questionnaire. If you fail to use this approach, respondents will become fed up with being exposed to questions that are irrelevant to them. They are more likely to not respond at all to the questionnaire. Again, it is a matter of making it easy for your respondents to do your bidding.

Pitfalls

There are many pitfalls and potential gaffs to be made in designing your questionnaire. Some will make your respondents smile, but most will make your questionnaire look unprofessional.

Language and wording

The wording of questions must be kept as straightforward as possible. Lengthy questions may lead to confusion and misunderstanding, as the reader needs to work at understanding them. Of course, there are occasions when an explanation is required, but these should be used sparingly and preferably in circumstances where a number of responses will follow—e.g. scales, ratings, and multi-response items so that the length of the question's preamble is in proportion to the volume of information gained.

The language and vocabulary should be appropriate to the intended reader. The use of jargon is to be generally avoided unless essential, or appropriate to the intended readership. If technical terms need to be used with a lay readership, you will need to explain them. The following is an example of two hypothetical questions in a health survey. The first is intended for medical practitioners, the second is for their patients.

- How many of your patients were born with an ASD, VSD or AVSD?
- To your knowledge, are you aware of any members of your family who were born with a 'hole in the heart' condition?

Leading questions

A leading question is one that gives the respondent the impression that you are looking for a particular response, or it gives a clear message about what are good or bad responses. The following is an example.

Q: Would you say that you spend too much on clothes?

Apart from the problem of determining what 'too much' is, the implication is that you **can** spend too much and that may be undesirable. Some feeling guilty about their spending, however large or small, might say YES, while others who regard themselves as moderate spenders will probably say NO. They may even feel insulted by the question!

A better question might be:

Q: On average, how much do you estimate you spend on clothes each month?

There are no socially value-laden messages in this question. It also allows you to quantify spending patterns, albeit an estimate.

Threatening questions

This does not necessarily refer to a physical threat, but more to questioning of or perceived challenge to a respondent's values, beliefs and behaviour. Any question that implies the respondent is somehow not operating in a socially acceptable way is unlikely to be well received! Embarrassing questions are also potentially threatening because they may be seen as exposing the respondent to ridicule or invasion of privacy. Any question that is likely to arouse anxiety in respondents should be regarded as *threatening*. Responses to such questions are likely to be biased in a way that neutralizes the *threat*.

Two questions in one

The respondent does not know how to respond because two or more responses become possible. The following is an example of this type of question.

Q: Substance abuse and violent crime are the most serious problems facing society today							
Strongly Agree		Agree		Not Sure		Disagree		Strongly Disagree	

A respondent may agree that drug abuse is the most serious problem facing society, but that violent crime is less so. How does the respondent answer this question? It requires a scale of agreement that is common to both issues, but the respondent may not have equality of agreement.

Ordering the questions

The use of funnel and inverted funnel sequencing of questions addresses one area of questionnaire structure. However, a questionnaire may be complex and cover a number of areas. In this case careful consideration needs to be given to the overall ordering of questions. The general rules are:

- Ask the easy to answer questions first.
- Closed response questions are often better placed before open response questions—they generally require less work to answer.
- Questions that are complex or relate to sensitive issues should be placed later in the questionnaire. This allows respondents to become comfortable with participating before being 'hit' with areas of difficulty. Also, a respondent will at least answer the early parts of the questionnaire

before refusing to continue. You may gain some useful data. A refusal to answer an early question is more likely to lead to refusal to complete the questionnaire.

The covering letter

If the questionnaire is part of a postal survey (as opposed to a survey carried out by an interviewer), it must be accompanied by a covering letter that explains its purpose and how the information will be used. In general potential respondents couldn't care less about your questionnaire, unless they have a vested interest in the subject. Where possible, you need to quickly establish the potential value of the survey to the reader. For example, you can indicate that the data will be used to improve services used by the reader.

A good covering letter should include the following elements.

- It should identify the person or organization carrying out the survey.
- It should explain the purpose of the survey.
- It should clearly explain why it is important for the reader to respond to the survey (i.e. address the reader's vested interest).
- It should address the issue of confidentiality. The reader must be confident that there is no potential to harm or embarrass. Of course, the requirements of the data protection act need to be observed. Will it be possible to identify an individual from any data published or passed on to a third party? These are important issues.

Finally, it is best practice to provide pre-paid return envelopes. It is unacceptable to expect a respondent to go to the expense of returning it.

Piloting the questionnaire

A new questionnaire must be piloted. It should be sent to a small sample of people and their responses assessed to check the suitability of each question as well as the overall reception. It is at this point that modifications can be made to remove problems in understanding questions, reducing ambiguity, identifying new categories of response, and generally tidy it up.

The pilot should be conducted under the same conditions as the real survey. Participants should be selected using the same random process as the real survey. It is difficult to be specific about the sample size for a pilot because this does depend on the overall size of the survey and the money and time available. Participants in the pilot should be excluded from the real survey.

Exercises

Activity 1

Develop a hypothesis about some aspect of **one** of the following areas:

1. The relationship between academic qualifications and employability;
2. The relationship between age and democratic participation;
3. The relationship between spatial awareness and sporting ability of professional athletes;
4. The relationship between academic ability and examination anxiety.

Activity 2

Develop your hypothesis from Activity 1 into a manageable and coherent research question.

Activity 3

1. Identify and define the concepts in your research question and operationalize them as research variables.
2. What were the problems and issues you had to address to arrive at *valid* and *reliable* variables?

Activity 4

Design a short questionnaire that will capture the data required to answer the research question. Pilot it with a few of your friends and evaluate its *reliability* and *validity*.

6

Measures of central tendency, variance and distribution

The most powerful statistical tests and analyses are those that are based on *interval* data (see chapter 1 for an explanation of *interval, ordinal* and *nominal* data). In this chapter, you will be introduced to some key properties of *interval* data variables that a number of tests of statistical significance rely upon (chapter 9 explains *statistical significance*). In chapters 9 to 11 you will be introduced to the most commonly used statistical techniques. By the end of this chapter you will:

- Understand and be able to calculate/determine mean, median, mode, quartiles and interquartile ranges, and their importance as descriptive statistics;
- Understand the principle of *normal distribution* as a theoretical model of variability in a population;
- Understand 'variance', 'standard deviation', 'z-scores' and 'standard error' as measures of dispersion and their significance in the social sciences;
- Be able to calculate measures of central tendency and variance by hand and with SPSS.

Measures of central tendency

One way of summarizing data is to determine a value that is typical of the sample as a whole. There are three main measures that are used. The *mean* and *median* values are important for a number of reasons. One reason is to tell us whether the data approximates the *normal distribution* pattern. The third measure, the *mode* is little used, although it can be useful.

Mean

To most people this means *average*. To use its correct title, it is called *arithmetic mean*, and it is calculated using this formula:

$$\bar{X} = \frac{\sum X}{N}$$

Where:

$\bar{X}$ = arithmetic mean

$\sum X$ = sum of all the values/observations

N = number of values/observations

Weighted mean

Almost always you will be dealing with data in which the frequency of a given value will be more than one. For example, Table 6.1 summarizes the ages of a sample of mature students and shows how the frequency of each is used to make calculating the *arithmetic mean* much easier.

Table 6.1:
Frequency table
of age

Age	Frequency	Frequency x Age
21	4	84
22	10	220
23	12	276
24	6	144
25	11	275
26	5	130
27	11	297
28	10	280
29	4	116
30	10	300
31	12	372
32	7	224
33	6	198
34	12	408
35	12	420

(Continued)

Age	Frequency	Frequency × Age
36	14	504
37	14	518
38	7	266
39	9	351
40	6	240
41	7	287
42	8	336
43	6	258
44	6	264
45	6	270
46	3	138
47	7	329
48	6	288
49	4	196
50	2	100
51	1	51
52	1	52
53	1	53
57	2	114
64	1	64
TOTALS	**243**	**8423**

◀ Table 6.1: (Continued)

Mean Age = 8423/243 = 34.7

Mean of grouped data

Where there are a large number of categories, some of which have very low frequencies, they may be 'collapsed' into a smaller number of categories by creating groups of equal intervals.

It is possible to calculate the mean value from those groups. You assume the typical value is the mid-point and multiply this by the group frequency to arrive at the weighted mean. Table 6.2 shows how this might be done using the data in Table 6.1 that is based on the response to the question: 'What was you age at your last birthday?'

Age group	Limits	Mid-point	Frequency	Mid-point × Frequency
21–26	21–25.9	23.5	43	1010.5
26–31	26–30.9	28.5	40	1140
31–36	31–35.9	33.5	49	1641.5
36–41	36–40.9	38.5	50	1925
41–46	41–45.9	43.5	33	1435.5
46–51	46–50.9	48.5	22	1067
51–56	51–55.9	53.5	3	160.5
56–61	56–60.9	58.5	2	117
61–66	61–65.9	63.5	1	63.5
TOTALS			243	8560.5

Arithmetic Mean = 8560.5/243 = 35.2

There is a small error introduced into the sum of **Mid-point × frequency** (8560.5 as opposed to 8423). The arithmetic mean is 35.2 years. The final result will be a close approximation. You simply need to recognize that there may be a relatively small error in the result. One way of overcoming that is to use groupings where the interval between minimum and maximum group value is quite small. An interval of 5 (e.g. '21–26') is likely to produce a smaller error than an interval of 10 (e.g. '21–31').

Problem with the mean

The mean is susceptible to distortion by extreme values. The *mean* is supposed to give us a representative value of a range of values. A simple example shows how this can be upset.

Example A: 2, 5, 6, 8, 11, **13**   MEAN = 7.5
Example B: 2, 5, 6, 8, 11, **55**   MEAN = 14.5

In Example B, there is an extreme value that is distorting, or skewing, the mean, making it unrepresentative of the sample. A mean of 7.5 is representative of Example A, but in Example B, a mean of 14.5 is skewed towards the extreme end, which does not make it truly representative.

Imagine that the numbers in both lists represented salaries in £10,000's in two very small companies (where the owner-manager had the highest salary). Both companies may want to represent themselves as offering very good salaries. Company B looks quite attractive with an average salary of £140,500, as opposed to £75,000 in company A.

There are two ways around this problem:

- The first is to identify *extreme values* and exclude them from the calculations. There is no problem so long as you are honest and indicate to the reader that the calculations have been based on adjusted figures.
- The second approach is to use a different measure for central tendency that is much less susceptible to this kind of distortion.

Median

The median is the measure of central tendency less likely to be subject to distortion by extreme values. The *median* is the middle value in a range of values. Again, this can best be illustrated with a simple example.

Example A: 2, 5, 6, 8 , 11, 13   MEDIAN = 7
Example B: 2, 5, 6, 8 , 11, 55   MEDIAN = 7

Where there is an odd number of cases, the median will always be the number in the middle—e.g. if there are seven cases, the median is the value of the fourth case. Where there is an even number of cases, as above, the median value lies between 6 and 8—i.e. 7, effectively the *mean* of 6 and 8. Here is another example where there is an even number of cases, but a wider gap between the middle two numbers

Example C: 2, 5, 6, 10 , 12, 15

The median lies between 6 and 10 (i.e. 8).

Another way of looking at the *median* is that it is the value where there are exactly 50% of cases below and above this value. That makes it important because it is fixed; you know where it is placed in your sample. The mean is not fixed like this.

It is often useful to calculate the *mean* and the *median* values for a variable and then compare them. If there is a large difference between them, this should be a warning that the values are skewed in some way by the presence of extreme values. In the examples used above, both values are very close—mean of 34.7 and median of 35.

Mode

This is of little value in statistical analysis. It is the most commonly occurring value within the data. In Table 6.1, there are two modal values, 36 and 37 years of age. The *mode* tells you where the peak of the distribution lies. Where *mode* does become interesting is if you detect two or more modal values that are not adjacent to each other, creating two peaks. It is just possible that this may be the result of your sample containing two different groups of individuals, each with its own mean and median, and it is worth exploring this with tests of significance (e.g. comparison of means, which will be explained later in the book).

Distribution and variance

Many statistical tests hinge upon the way data is distributed. This section of the chapter looks at the two properties of data frequency distribution and the spread or *variance* of data.

Normal distribution

Difference is what allows us to differentiate between objects and people. The really interesting thing about difference is that it follows rules—or at least patterns—of a sort. One of these patterns of frequency distribution is called the *normal distribution* or *Gaussian distribution*, named after the early nineteenth century German mathematician, Karl Gauss.

The distribution and its meaning can be illustrated with a simple, but unlikely, example. Take all blades of grass in the world, for example. In that population we can observe an obvious wide range of lengths, so that if we were to measure the length (a *continuous interval variable*—see chapter 1 to

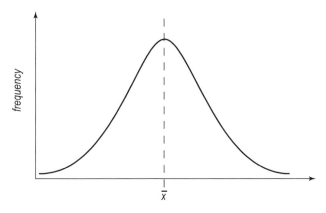

continuous variable being measured

Figure 6.1: The normal distribution curve

remind you of the definition of this) of every blade of grass in the world and plotted length against the frequency of each length (i.e. the number of occurrences) we would almost certainly observe the normal distribution curve such as that shown in fig 6.1.

This bell-shaped curve seen in Fig 6.1 is the theoretical result of plotting the distribution of an interval measure in a population, and it has a number of key properties that are important in the use of statistical techniques. These properties are:

1. the curve is based on an infinite number of observations/ measurements[1]
2. it is symmetrical and bell-shaped;
3. the mode, median and mean coincide at the centre of the distribution—line of symmetry—perfect symmetry must mean these are all the same value;
4. and 50% of observations lie below the line of symmetry and 50% lie above it;

---

[1] While there are a finite number of blades of grass on the planet, the number is so impossibly large it is a good approximation to infinite in the practical sense.

5. there is a direct mathematical relationship between the frequencies and the values of the variable.

It is a theoretical concept but a number of tests of statistical significance discussed in later in this book do depend on the extent to which your research observations approximate the *normal distribution.*

- *Parametric tests* are used where the data distribution approximates or is assumed to approximate the normal distribution;
- *Non-parametric tests* are used where the data does not approximate the normal distribution.

This distinction is extremely important when selecting appropriate statistical techniques to test your research findings, and these will be covered in some detail. For now, the remainder of this chapter concentrates on several statistical measures of the distribution—or *variance*—of your data.

Range, minimum and maximum values

The *minimum* and *maximum* values of any data set are self explanatory—i.e. the lowest and highest respectively. The *range* is quite simply the difference between these two values. In **Table 6.1**, 21 and 64 are the minimum and maximum values of mature student ages, with a range of 43 years. Knowing these three values gives us a clear idea of the spread of data. This can be quite important as it helps us decide if our sampling technique has worked. So if the age range in the sample does not approximate that of the research population (e.g. the sample age range may be very much smaller than the population age range purely because of sampling error), we may need to be very careful about extrapolating our results from the sample to the population.

Another example of this is in psychological tests or school examinations. A good test is effective in discriminating between individuals taking the test and if we assume that natural ability, for example, varies amongst the population in a way that approximates the normal distribution, there should be a very wide range of performance in the test to reflect this.

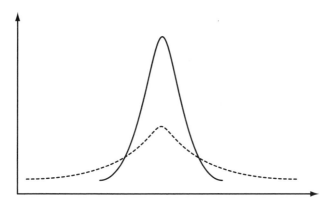

Figure 6.2: The normal distribution curve may take many shapes

So, if the range of scores in a test is between 10% and 95%, it would seem at first sight to be an effective discriminator—much more so than a test whose range of scores is from 65% to 85%. The latter range is so narrow we have to question the effectiveness of the test as this does not appear to reflect the natural range in the population as a whole.

Variance and standard deviation

The assumption of data approximating *normal distribution* is important in statistics as it forms the basis of what are termed *parametric tests* of *statistical significance*[2]. We tend to use *standard deviation* more frequently than *variance*. Before looking at the mathematics behind these two measures of data dispersal, it is useful to return to the *normal distribution curve* to explore it in more detail. The curve may take any number of forms and fig 6.2 shows two curves superimposed on each other to show how there is a general form, but can be different shapes.

As fig 6.2 shows, while the overall form of symmetrical distribution of data around the mean is always the same, the data may be more or less dispersed as these two curves show.

---

[2] The concept of statistical significance was introduced in chapter 3 and is dealt with in more detail in chapter 9.

The solid line curve shows the data bunched up around the mean, with a smaller range between minimum and maximum values. The broken line curve is much flatter with a greater range and so more widely dispersed data about the mean. If these curves represented the distribution of scores on two test papers, for example, the broken lined curve shows greater ability to discriminate between candidates sitting the test—it shows up differenced more clearly that the solid lined curve. The broken lined curve is showing greater *variance* and *standard deviation.*

Skewed distribution

Not all interval data variables will show a normal distribution—the data may be clustered at one end of the distribution, with the rest of the data training off as a long tail. Fig 6.3 shows the two types of *skew.*

Fig 6.3 shows the effect *skew* has on the relationship between the mean and median. The mean and median are

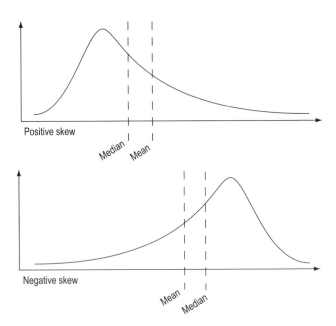

Figure 6.3: Positive and negative skew

identical in data that is normally distributed because of its symmetry. The clustering of data to the left (*positive skew*) means that the median value is lower than the mean because the greatest frequencies of the data are in the lower value region. In *negative skew*, where data is clustered to the right, the reverse is the case: the median is greater than the mean. A visual examination of the median and mean values helps to alert the researcher to possible skew.

Tests of statistical significance that assume normal distribution (referred to as *parametric tests*) would not normally be employed. This will be discussed further this in chapter 9.

Standard deviation

Standard deviation enables us to make direct comparisons because it is a standardized measure of data dispersion—i.e. spread. It assumes normal distribution of the data or at least nearly so) and it has significant and powerful properties illustrated by fig 6.4.

Standard deviation is effectively the average deviation of values from the *mean*. The curious thing about this value is that there are a fixed percentage of cases that lie within

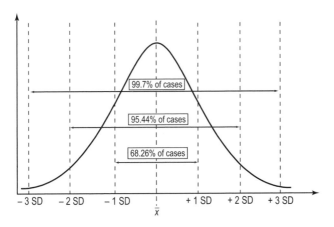

Figure 6.4: Standard deviation and distribution of cases

the range from the value of the mean and one standard deviation from it:

- 68.26% of cases will lie between the range of +1 and −1 standard deviation (i.e. 34.13% of values lie either side of the mean)
- 95.44% of cases lie between the range of +2 and −2 standard deviations, and
- 99.7% of cases lie between the range of +3 and −3 standard deviations.

These are properties of any normally distributed data and so is a good indicator of how dispersed the data is. Looking at fig 6.2, the standard deviation for the solid lined distribution will be less than that for the broken lined distribution—the data is more tightly bunched. Before looking at how to calculate the standard deviation of any data set, look at a practical example of the use of the measure. Table 6.3 compares the ages of mature students at two colleges.

	College X	College Y
Mean	34.7	35.0
Median	35.0	34.8
Standard deviation	8.4	4.3

Table 6.3: Comparison of student populations in two colleges

Table 6.3 provides the following insights into the age distributions:

- The mean and median values are very close together, suggesting a normal distribution of age data (or very nearly so).
- The standard deviation of students' ages at college X is almost twice that of college Y—suggesting a greater diversity of age range at college X:

Using this summary data, we can make a number of important estimates about the two colleges, and these are shown in Table 6.4. If you were to plot the data as a graph it would broadly resemble the form shown in fig 6.2.

	College X	College Y
Mean	34.7	35.0
Median	35.0	34.8
Standard deviation	8.4	4.3
Approx 68% of students are aged between	26 and 43	31 and 39
Approx 95% of students are aged between	18 and 52	26 and 44

◄ **Table 6.4:** Confidence limits

The estimates in Table 6.4 are a little rough and ready because fractional ages have been rounded up or down, but the data clearly shows how knowing the standard deviation helps us understand an important difference between the age distributions of the two colleges. If these were examination scores rather than ages, we would be able to see that the exam at college X was more effective at differentiating student ability levels (this does, of course assume the ability range in both colleges is more or less the same—yet another assumption!).

Calculating standard deviation

These days, this kind of calculation will be left to SPSS or MINITAB, but it is as well to have a working knowledge of the method for some of the more basic measures. The formula for calculating standard deviation is:

$$s = \sqrt{\frac{\sum (X - \bar{X})^2}{N}}$$

Translated this means:

s = standard deviation

$\sqrt{\ }$ = square root

$\sum$ = sum of (total)

$(X - \bar{X})^2$ = squaring the difference between a value (e.g. the score of a college student) and the mean of the variable (e.g. the mean score of college students)

N = the total number of cases in the sample.

An example will best explain the process.

Five students score the following percentages in a mathematics exam: 72, 81, 86, 69, and 57.

X	$(X - \bar{X})$	$(X - \bar{X})^2$
72	(i.e. 72–73)–1	1
81	8	64
86	14	169
69	–4	16
57	–16	256
Mean $\bar{X}$ = 73	$\sum(X - \bar{X})^2 =$	506

Putting these values into the SD equation:

$$s = \sqrt{\frac{506}{5}} = \sqrt{101.2} = 10.1 \text{(rounded to 1 decimal place)}$$

Working with a small sample like this makes the manual calculation pretty straightforward, but with a large survey where there may be hundreds of cases, this is a nightmare. Let SPSS take the strain. Also, with a small sample, the value of standard deviation is very much diminished to the point of being meaningless.

The use of SPSS to calculate standard deviation is explained towards the end of the chapter.

Z-score

You will come across this occasionally in statistical work, and a number of statistical tests are based on the *z-score* and it is simply the number of *standard deviations* a particular value lies above or below the mean (i.e. it is directional so may be + or –).

For example, in an examination, the mean score is 45% and the standard deviation is 12. If a student achieves a score of 72%, his or her *z-score* is:

$$z = \frac{72 - 45}{12} = \frac{27}{6} = 2.25$$

The student's score is therefore at the top end of the range of marks (remember that 95.4% of all scores will lie within ±2 standard deviations from the mean

The general formula for z is:

$z = (score - mean) \div standard\ deviation$

So, if a student achieves 35%, the *z-score* will be:

$$z = \frac{35 - 45}{12} = \frac{-10}{12} = -0.83$$

In other words, the *z-score* is negative as it falls below the mean value.

Variance

A number of tests for statistical significance also rely on *variance*. This is very closely related to standard deviation and SPSS printouts often include a measure of *variance*.

Stripped to its essential feature, *variance* is also a measure of the dispersion of the data—i.e. a measure of the spread of data and is calculated by squaring the standard deviation. So, for example, the data on the college students' age ranges do show quite a large variability between the two colleges (Tables 6.3 and 6.4):

- Variance college X = $8.4^2 = 70.6$
- Variance college Y = $4.3^2 = 18.5$

What this is showing is that while the mean age of students in each college is very similar, the *variance* is very large and so there does appear to be significant difference between the two populations—and this will be explored in Part 2 of the book.

Percentile, deciles and quartiles

These measures are directly linked to the *median*. Deciles and quartiles are frequently used in social research to make comparisons linked to social conditions and social policy. All of these measures can be calculated using SPSS.

Percentiles and deciles

The median is the 50% point in the distribution of cases. In other words it is the 50th percentile—there are 50% of cases above and 50% of cases below this point. If the distribution of cases is divided into 100 equal frequencies, then each is a percentile.

This is often used in ranking test scores in psychometric assessment. So, for example if an individual's test score is at the 95th percentile, this means that only 50% of people scored higher. Put another way, the individual did better than 95% of all those who sat the test.

If the data frequencies are divided into 10 equal portions, then each portion is a *decile*.

Quartiles and interquartile range

Another measure used to describe the spread of data is based on the *median* value as the point of centrality rather than the *mean*. It uses three fixed points based on the numbers of cases below that particular point. The fixed points are called *quartiles*. The *first quartile* is the value that represents the point where 25% of cases lie below it. The *third quartile* is the value that represents the point where 75% of cases lie below it. The second quartile is, of course, the *median*. The difference in value between the *first* and *third quartile* is called the *interquartile range*.

It has the advantage of not being affected by extreme values because it is not directly reliant upon the values themselves, but on the number of cases. This makes it valuable in making direct comparisons between groups. Comparing the *interquartile range* and *median* helps overcome the problem described earlier by *extreme values*.

Standard error of measurement (SE)

You must remember that you will almost always be working with a sample drawn from a population rather than the entire population. All the statistical processes used in social science research relate to the sample, not the population as a whole. Since a sample may not be a perfect representation of the population, there are problems when you extrapolate results to the population as a whole.

The *standard error of measurement* helps get around that problem. The idea is as follows. If you draw a random sample from your population and calculate a statistical measure, you will arrive at a set of values. If you now repeated the process of randomly drawing your sample from the whole population, the membership of the second sample would be different to the first. Of course, there may be some of the original sample members there, but it would still be different. After calculating your statistic again it may be a different value. For example, the mean age of the first sample may have been 32 and that of the second was 34. If you repeated this process many times, you would expect to have a range of values for your mean age. If you plotted a graph of these results you would expect a normal distribution, with clustering around the mean of the means.

This mean of all the sampling means is likely to be the true mean of the population. However, it could be any of the other values! If you calculate the *standard deviation* of these values you can create *limits of confidence*. For example, if the mean of your sample measures is 30, and the standard deviation is 2, you can be *68% confident* that the real value lies between 28 and 32, and *95% confident* that it lies between 26 and 34.

This special case of *standard deviation* is called the *standard error of measurement*. It recognizes that all measurements are subject to error and is a way of quantifying the error. SPSS and MINITAB automatically calculates the SE alongside other tests.

Calculating standard error of measurement (SE)

SPSS and MINITAB will calculate the standard error, but if you recall in chapter 4 on probability sampling, you can work backwards from acceptable sampling error to the sample size. This is the formula you would use to determine SE, or to calculate sample size from what you consider to be an acceptable margin of error.

$$SE = \frac{s}{\sqrt{n}}$$

Where:

$s =$ the standard deviation of the sample variable being measured

$n =$ sample size

If you are working backwards to determine the sample size, then both SE and s have to be estimated either from experience, or what you would consider an acceptable margin.

If the population is small, a *finite population correction* must be used, modifying the formula:

$$SE = \frac{s^2}{n}\left(1 - \frac{n}{N}\right)$$

Where $N =$ population size.

SPSS tutorial

Calculating measures of central tendency and dispersion

Q: Explore the salaries of employees in a large technology-based company.

This worked example will show how SPSS may be used to calculate and display these descriptive statistics. The data used has been drawn from a study of salary levels in an organization.

1. Access to the statistics needed is through the **Analyze** > **Descriptive Statistics** > **Explore** (fig 6.5).

2. The variable *Current Salary* is selected and transferred to the Dependent List by clicking the arrow button (circled), click the statistic checkbox (circled) as this will generate descriptive statistics, and then click the Statistics button (circled) to select the measures (fig 6.6). For the purpose of this example, only descriptives are required—the 95% confidence limit is the default setting, but it can be changed to 99%. Click Continue to

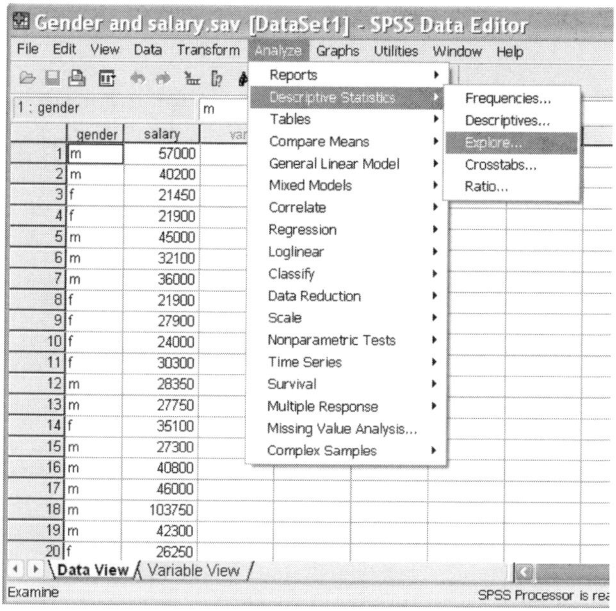

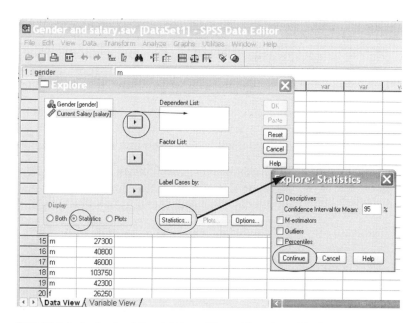

Figure 6.5: Menu route to the basic exploratory statistical indicators

Figure 6.6: Selecting the descriptive statistics and the confidence interval for the mean

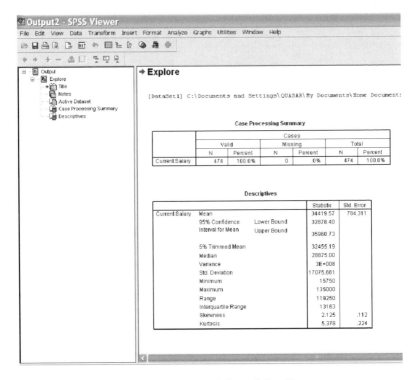

Figure 6.7: SPSS Output window showing the descriptive statistics table

close the Explore: Statistics box, then OK in the Explore box to start the calculations.

3. The output screen of SPSS presents the results of its calculations (fig 6.7). So, for this data we can see that:

   a. There are 474 cases in the database.

   b. The mean salary of the sample group is £34419, but since the standard error is £784, we are 95% confident that the population mean lies between £32878 and £35960 (i.e. ±2 standard errors).

   c. The median, minimum and maximum salaries are £28875, £15750 and £135000 respectively.

   d. The median value is lower than the mean, so there may be a small amount of positive skew in the distribution.

Exercises

Progress questions

1. Explain the difference between: (a) mean and median (b) standard deviation and interquartile range.
2. If the *mean* score in a survey of attitudes towards televised sport is 70 and the standard deviation is 20, what range of scores would fall within the (a) 95% confidence limits; (b) 99% confidence limits?
3. Why are the *median* and *interquartile range* not affected by extreme values?
4. The mean age of a sample of people is 25. The standard error is 1.5. Using the 95% confidence limit, what range is the actual mean age of the population likely to be?
5. Calculate the standard error in the mean monthly consumption of chocolate bars in a population if the sample size is 900 and the standard deviation of the number of bars consumed each month is 3.

Assignment

Using the database you set up during the Assignment for chapter 2 based on the Excel file satisfaction.xls and its coding frame (coding_frame.pdf), generate *means, medians, standard deviations, standard error* and *interquartile ranges* for all the *interval* variables. Look at each of these measures and comment on the extent to which the distributions appear to approximate *normal distribution*. Also, use the standard error of measurement for each of these and determine the 95% confidence limits for these variables.

7 Indices, scales, weighting and ratings

Learning outcomes:

Official government statistics as well as academic social research data is often presented using *indices*, *scales* and *ratings*. By the end of this chapter, you will:

- Be able to define and calculate indices;
- Understand the role of indices in statistics;
- Know what is meant by 'weighting' data;
- Understand why 'weighting' is used and be able to apply weighting appropriately to datasets.

Overview

An *index* is a useful way of making comparisons of values over time or across geographical areas or social groupings. Actual quantities are not needed so the problem of dealing with large cumbersome numbers is eliminated. An *index* represents a value as a proportion of another (effectively as a percentage).

A scale is a means of measuring differences between individuals. For example, a scale may measure differences in attitudes towards a concept. *Likert scales* do this very effectively by asking respondents to indicate their level of agreement with statements. *Semantic differentials* require respondents to indicate their reaction to bi-polar descriptors (e.g. Good/Bad) that places them on a continuum from one extreme to the other.

Ratings enable the researcher to measure a respondent's perception of the quality of experience, while *ranking* gains a respondent's views on the relative importance of individual statements within a group of statements.

Indices

An *index* represents one value as a proportion of another. Indices are frequently used where a standardised means of comparing values is needed. It is particularly useful for comparing changes over a period of time or for showing differences between individual subjects of your study (e.g. people or geographical regions). **Table 7.1** shows a typical example of where indices are used in official government statistics.

Table 7.1:
Economic
statistics 2003

Economic statistics, 2003					
	People in employment (thousands)	Percentage employment in			Gross domestic product per head EUR 25 =100, 2002
		Agriculture	Industry	Services	
EUR 25	193,221	–	–	–	100.0
Austria	3,736	5.6	29.3	65.1	120.8
Belgium	4,070	1.8	24.8	73.4	116.8
Cyprus	327	5.2	22.9	71.9	82.9
Czech Republic	4,701	4.5	39.6	55.8	67.6
Denmark	2,707	3.2	23.8	72.9	122.5
Estonia	594	6.2	32.5	61.3	46.6
Finland	2,365	5.1	26.2	68.3	113.8
France	24,584	4.2	24.4	69.0	113.0
Germany	35,927	2.4	31.4	66.2	108.7
Greece	4,274	15.4	22.6	62.0	77.6
Hungary	3,922	5.3	33.4	61.3	58.6
Ireland	1,797	6.5	27.6	65.5	132.7
Italy	22,054	4.9	31.8	63.3	109.0
Latvia	1,007	13.7	27.0	59.3	39.0
Lithuania	1,433	17.9	28.1	53.9	42.4
Luxembourg	188	–	19.3	78.1	212.7
Malta	148	2.2	29.9	67.9	73.2
Netherlands	8,121	3.0	20.8	76.2	122.1
Poland	13,617	18.4	28.6	53.0	45.6

(Continued)

Economic statistics, 2003				
	People in employment (thousands)	Percentage employment in		Gross domestic product per head EUR 25 =100, 2002
		Agriculture	Industry	Services	
Portugal	5,118	12.5	32.3	55.2	76.7
Slovakia	2,162	5.8	38.3	55.8	51.3
Slovenia	897	8.4	37.5	53.4	75.3
Spain	16,695	5.6	30.6	63.8	94.6
Sweden	4,314	2.5	22.5	74.9	114.8
United Kingdom	28,696	1.3	23.3	75.2	117.8

◀
Table 7.1:
(Continued)

Source: Eurostat; office for national statistics

The table summarizes a number of key economic statistics comparing the data on the EU states in 2003. The final column is an *index* and provides a way of comparing performance in a standardized way that is based on proportion rather than a direct comparison of raw values.

The index compares the Gross Domestic Product (GDP) per head of population. Clearly the population of states varies and so the use of an *index* makes the comparison independent of numbers. An *index* requires the use of a base value and in this case, this is taken as the mean GDP of all EU states—this is recorded as 100. In effect, the index represents a percentage of the base value and provides a visual comparative measure—in this case a measure of relative national wealth. From this table it can be seen that in 2003, Latvia has the lowest national wealth at 39% of the EU average, and Luxembourg the highest at 212.7% of the EU average.

Creating an index

Creating an *index* is a straightforward percentages calculation, but some thought is needed to identify its purpose. What are you comparing? There can be much debate around this area of index construction, especially where social policy is concerned. The consumer price index (CPI), for example, is a measure of consumer spending, and this value is compared with the value twelve months ago. If the value of average consumer spending in July 2007 was £756 and its value in July 2006 was £740 the CPI would have been:

$$CPI = \left(\frac{Value\ in\ 2007}{Value\ in\ 2006} \right) \times 100\%$$

$$CPI = \left(\frac{756}{740} \right) \times 100 = 102.2$$

This would make the annual inflation rate in July 2007 as 2.2%.

The general calculation for creating an index is:

$$\text{Index} = \frac{\text{Value to compare}}{\text{Base value}} \times 100$$

Indices are interval measures.

Weighting an index

An *index* based purely on raw values of its components can present some difficulties because it does not show the relative influence of each component. For example, suppose you want to create an index of individual income as a measure of personal affluence. The problem you face when comparing different areas will be their population structure. In general, retired people tend to have a lower income than younger working people. Also people in their late 30's, 40's and 50's generally have a higher income than those in their early 20's. In other words, there can be influences on the income that you will want to control for.

You can *weight* the index so that like is compared with like, so no matter what the age group composition of an area you can make a comparison that is independent of this potential bias. **Table 7.2** shows hypothetical data on mean incomes for the UK and a seaside town on the south coast.

There are two features of note. The seaside town has a higher proportion of elderly people who appear more affluent than the hypothetical UK elderly population. It also has an under-60 population who appear to be less affluent than the hypothetical UK population. An index based on the mean of means using the UK as the base is calculated as follows:

Age group	UK (1994=100)		Seaside town	
	% Age group*	Mean income	% Age group*	Mean income
16–44	42	£350	30	£300
45–59	20	£400	25	£350
60–79	14	£185	30	£200
80+	4	£160	10	£200
Mean		£273.75		£262.50

Table 7.2: Hypothetical data on mean lincomes for UK and a seaside town

* Under-16s have been excluded, so % do not total 100%.

$$Income\ Index = \frac{Mean\ Income\ seaside\ town}{Mean\ Income\ UK} \times 100$$

$$Income\ Index = \left(\frac{262.5}{273.75}\right) \times 100 = 95.9$$

In other words, the population of the seaside town appear to be less affluent than the UK population as a whole. While this is fine as far as it goes, there may be difficulties later when attempting to make comparisons over time. How can you tell that changes are due to real changes in income or changes in the age distribution of the population?

You can overcome this by standardising the age distribution by making the calculation assuming the UK age group distribution for the seaside town. **Table 7.3** shows this.

Age group	UK (1994 = 100)			Seaside town	
	% Age group*	Mean income	Weighted aggregate*	Mean income	Weighted aggregate
16–44	42	£350	£14,700	£300	£12,600
45–59	20	£400	£8000	£350	£7500

Table 7.3: Age adjusted index

(Continued)

Table 7.3: (Continued) ▶

Age group	UK (1994 = 100)			Seaside town	
	% Age group*	Mean income	Weighted aggregate*	Mean income	Weighted aggregate
60–79	14	£185	£2590	£200	£2800
80+	4	£160	£640	£200	£800
		TOTALS	**£25,930.00**		**£23,700.00**

= % Age Group x Mean Income.

Adjusted Income Index = (23,700/25,900) × 100 = 89.5

The new figure reduces the impact of the affluent elderly population that has distorted the picture and plays up the effect of the younger population in keeping with the rest of the UK. This difference is quite important, as the way age groups spend or save their income is often quite different.

Changing the index base

It is sometimes necessary to change the base of your index to make new comparisons. **Table 7.4** is a hypothetical table of indices with 1990 as the base used by a company to calculate special bonuses. If the company decided to revise this scheme in 1993 so that future bonuses would be calculated from the *index* using the 1993 value as its base because of new working practices, the *indices* would have to be recalculated. The table shows the old and new index values. The new value for 1990 would be calculated like this:

$$\text{New Index} = \frac{\text{Old Index Value}}{\text{New Base}} \times 100$$
$$= (100/95) \times 100 = 105.3$$

Table 7.4: recalculating indices with a new base ▶

Year	Value of old index (1990 = 100)	Value of new index (1993 = 100)
1990	100	105.3
1991	95	100
1992	90	94.7

(Continued)

Year	Value of old index (1990 = 100)	Value of new index (1993 = 100)
1993	95	100
1994	102	107.4
1995	110	115.8
1996	115	121
1997	132	138.9
1998	130	136.8
1999	140	147.4

◀
Table 7.4:
(Continued)

Scales

You may want to measure the strength of a respondent's views on matters of interest, opinions or attitudes. You can use a *scale* to be able to differentiate between respondents so that it is possible to say respondent 'A' feels more strongly about subject 'X' than respondent 'B' does.

A *scale* is an *ordinal* measure and so should not in theory be subjected to statistical techniques that assume *interval* measures and *normal distribution* of data. However, there is an exception to this restriction when the scale scores are aggregated. This will be described a little later.

Likert scale

A Likert scale is often used to measure respondents' attitudes or opinions. A question would be in the form of a statement to which respondents are asked to indicate their level of agreement. For example, the following question asks respondents to indicate their level of agreement with a statement about professional development and training by ticking the relevant box:

Q: I believe it is important to keep my skills and knowledge up to date.

| Strongly agree | Agree | Not sure | Disagree | Strongly disagree |

The responses are coded 1 to 5 with 1 for **strongly disagree**. The five-point scale is important as it allows for a neutral response. A three-point scale (i.e. agree, not sure, disagree) would do, but you would be less able to say something about the strength of agreement or disagreement and is not so sensitive a measure. However, to create a *scale,* you will need to identify a number of attitudinal statements that reflect a particular factor you are investigating and then generate an aggregate score for each individual. For example, continuing the theme of worker attitudes to continuing professional development, the following looks at some of the items that may be incorporated into a scale.

	Strongly agree	Agree	Not sure	Disagree	Strongly disagree
1. *I believe it is important to keep my skills and knowledge up to date.*					
2. *If I find that my knowledge and skill need updating, I request training.*					
3. *I believe it is my employer's responsibility to identify my training needs, not mine.*					

Item 3 can be regarded as a negative item as agreement with this item may be considered as implying a negative attitude by denying personal responsibility. In this case, the coding would be reversed (i.e. Strongly Agree would be coded as 1 and Strongly Disagree coded as 5) to reflect the need to score positive attitudes highly.

Each respondent's score is obtained by totalling the scores for each item. For example, the maximum score on the three questions above would be 15, with 3 as the minimum. There is, therefore a 12-point range of scores along which each respondent may lie. The aim is to be able to differentiate between individuals with respect to their attitudes towards professional development. The more items there are in a scale, the greater the range of scores and so the more discriminating the scale will become. Under these circumstances it is possible to treat a *scale* as an *interval* measure.

If you plan to construct a scale for use on many occasions, it is important to establish its ability to discriminate between individual respondents.

Discriminative power (DP) of a Likert scale item

The ability of a scale to differentiate between individuals by creating a wide range of scores along which your respondents are distributed is referred to as its *discriminative power*, or DP for short. If the range of scores is relatively narrow, respondents will tend to cluster on the scale, making it ineffective because you cannot easily differentiate between individuals.

The process of constructing a scale based on good DP is:

1. Devise as many items as possible that could be used to test a respondent's attitude to the subject under consideration (e.g. attitude to professional development).
2. Pilot the items by administering them all to a sample of people drawn using probability sampling.
3. Score each respondent's responses.
4. Calculate the DP of each question.
5. Construct the scale using items with the highest DP values.

The DP for each question is calculated by taking the highest 25% of scores and the lowest 25% of scores (first and lowest quartiles) to firstly determine mean score in each of the two quartiles. The DP is then calculated by subtracting the two means. The greater the difference (i.e. the higher the DP) the more effective it will be as a measure.

Table 7.5 shows a hypothetical example of this type of calculation.

	No of respondents	Frequency of scores					Weighted aggregate	Weighted mean
		1	2	3	4	5		
1st Quartile	18	0	2	4	6	6	70	3.9
4th Quartile	18	2	16	0	0	0	34	1.9

The DP of this scale item is 3.9–1.9 = 2.

Semantic differential

At first sight this scale may look like a Likert scale, but there is a fundamental difference. While you will be asking respondents to record their reactions, it is based on their reactions to a concept using a bi-polar scale. It is quite a good way of obtaining a rating of a particular concept or activity. The two ends of the scale item use contrasting adjectives. Whereas a Likert scale item will only ask for the degree of agreement with a statement, the *semantic differential* presents a continuum and asks respondents to place themselves on that continuum.

A simple example of a *semantic differential* item is given below.

(The scale shows a continuum from one extreme to the other)							
	Very	Fairly	Slightly	Neither	Slightly	Fairly	Very	
Good								Bad

Respondents will record their reactions ranging from very good to very bad. A *semantic differential* will provide you with rich descriptive data on any particular activity or object. The responses are coded using a 7-point scale with 1 and 7 representing the opposite ends of the continuum, and 4 representing the neutral position.

Below is a hypothetical example of a *semantic differential* in which respondents are asked to provide their reactions to descriptions of MPs.

Q: Here is a list of pairs of words that could be used to describe Members of Parliament. Between each pair is a set of boxes. Taking the first pair of words Truthful/Untruthful, the box on the extreme left means that you believe MPs are very truthful, while the box at the far right means that you believe MPs are very untruthful. Taking each pair in turn, please tick the box that best describes your view of MPs.

	Very	Fairly	Slightly	Neither	Slightly	Fairly	Very	
Truthful								Untruthful
Hardworking								Lazy
Well-informed								Ill-informed

Rating and ranking

Rating questions generate information about the quality of a respondent's experience of an event, activity or concept. *Ranking* questions generate information about a respondent's view on the relative importance of items in a list.

Rating

The use of *rating* questions is very common in evaluation questionnaires, or market research. A rating response is an ordinal measure because you cannot quantify **Good** or **Very Good**, for example. A hypothetical example of a rating question is given below:

Q: Please rate the quality of the service you received by placing a tick in the appropriate box for each of the factors listed. For example, if you rate the quality of the nursing care as Very Good, tick the Very Good box.

	Very Good	Good	Satisfactory	Bad	Poor
Quality of the nursing care.					
Willingness of the medical staff to listen to my concerns.					
The information given to me about what my operation involved.					

Each of the responses may be coded 1 (for **Poor**) to 5 (for **Very Good**).

Ranking

Many concepts in social science cannot be quantified—even at the **Very Good** to **Poor** level of rating—but it is possible to determine the level of importance as far as an individual respondent is concerned. During analysis later, it may be possible to find patterns of ordering among sub-groups of respondents.

Below is an example of a *ranking* question.

Q: The three statements below refer to lifestyle characteristics that people consider important for their family. Please examine all three statements and decide which <u>one</u> is most important to you. Circle the figure 1 next to the statement. Circle the figure 3 against the statement that is least important to you. Circle the figure 2 against the remaining statement.

A high family income.	1	2	3
Time spent together in family activities.	1	2	3
My career ambitions.	1	2	3

An important feature of *ranking* questions is the element of forced choice. The respondent is being asked to think very carefully about what is important and then make choices about relative importance. Unlike scales and ratings, it is not possible to have the same response for each item. While very useful, these questions are often very hard for a respondent who may find some conflict between competing alternatives! For example, a high family income may be very important because it represents financial security, but the respondent has to consider if that is more important than spending time with the family. The relative order of a respondent's choice may help to explain behaviour and attitudes.

Exercises

1. Why would you use an index to compare the relative affluence of various places in the UK?
2. What kind of information are Likert scales good at collecting?
3. Which type of question would be good at finding out the relative importance of a set of items covering judgements about what constituted a 'good life', a rating question or a ranking question?
4. What type of information is a semantic differential good at collecting?
5. Design a *Likert scale* to measure respondents' attitudes to an issue of importance to you (e.g. a local planning decision or environmental issue)
6. Design a *semantic differential* scale to measure respondents' reactions to a new ready-prepared meal that is soon to be released on the market.
7. Design a rating question to identify respondents' view of a local public service.

8 Initial exploration of your data

Learning outcomes:

The data collection stage of your research has been completed, and the data recorded in SPSS. Now the work of making sense of the data has arrived. Often there is a great deal of it: a simple survey of say 100 people gathering data on ten variables produces 1000 individual data items—i.e. 1000 data cells in the database. There is scope for error in entering the data, but you will also want to scan the data for obvious anomalies or patterns. For example, are interval data variables showing a normal distribution, or pretty close to it? Are there individual values that seem off the wall—could be a data entry error, sampling error or a respondent playing games and not being entirely honest—it happens!

By the end of this chapter you will:

- Understand that visual presentation of data is valuable for initial exploration of data to identify distribution patterns;
- Understand how frequency distributions determine the way in which statistical tests are used on the data—i.e. parametric and non-parametric tests of significance;
- Be able to identify appropriate visual methods of representing and exploring data;
- Be able to interpret the results of the presentation methods listed above;
- Be able to use SPSS to carry out initial exploration of data.

Univariate analysis

When you have collected your data and entered it into a database or statistics software package, you will want to begin the process of analysis. The first stage is to look at each variable individually. This is called *univariate analysis*. There is little in the way of detailed analysis you can do when looking at the data for a single variable, but it is useful for the following.

- You are able to identify *rogue values* and *outliers*. See **Cleaning Up the Data Set** later in this chapter.
- You can look at the general distribution of the data to assess its spread. Is it all bunched up across a narrow range of values, or is it very widely dispersed? Does the pattern of distribution of *interval* data approximate *normal distribution*? (Remember some statistical techniques require the data to approximate a normal distribution pattern).
- You can calculate some simple statistical *descriptives* that enable you to make some comparisons—these include *mean, median, standard deviation* and *interquartile range.*

The chapter also explains how SPSS is used to explore data and generate the tables, summary statistics and charts. Of course, this can all be done by hand as well, but let the computer take the strain on this one.

Frequency tables

The first task after completing the data entry stage of your project is to generate a list of responses to each variable in the database. *Frequency tables* are generated. This will list the number of responses to each option (item) within a question.

Generating a frequency table

All spreadsheet and statistics software packages will generate frequency tables. Table 8.1 shows a frequency table generated by SPSS. Age of respondents is an interval measure and so we would also want to produce the summary statistics for this value as well: i.e. mean, median, standard deviation, standard error and possibly quartiles and interquartile range. These will give some additional information about the dispersion of the data (Table 8.2).

The statistics in Table 8.2 give us some insight into the nature of the data's distribution. For example, the median and mean values are very close to each other, suggesting the possibility of a normal distribution—i.e. symmetry around the mean. At worst there seems to be a very slight negative skew. Another clue is shown in the percentiles section. There is something close to symmetry around the median value. The range between the first quartile and the median (i.e. the second quartile) is seven years, and between the median and the third quartile is six years. These are very similar values hinting at a symmetrical distribution—i.e. normal distribution. As you will see later in the chapter, the data should also be represented graphically to gain a visual confirmation of the distribution.

Q1 Age	Age	Frequency	Percent	Valid percent	Cumulative percent
Valid	21	4	1.6	1.6	1.6
	22	10	4.1	4.1	5.8
	23	12	4.9	4.9	10.7
	24	6	2.4	2.5	13.2
	25	11	4.5	4.5	17.7
	26	5	2.0	2.1	19.8
	27	11	4.5	4.5	24.3
	28	10	4.1	4.1	28.4
	29	4	1.6	1.6	30.0
	30	10	4.1	4.1	34.2
	31	12	4.9	4.9	39.1
	32	7	2.8	2.9	42.0
	33	6	2.4	2.5	44.4
	34	12	4.9	4.9	49.4
	35	12	4.9	4.9	54.3
	36	14	5.7	5.8	60.1
	37	14	5.7	5.8	65.8
	38	7	2.8	2.9	68.7
	39	9	3.7	3.7	72.4
	40	6	2.4	2.5	74.9
	41	7	2.8	2.9	77.8

◄ Table 8.1: Frequency table of age of respondents

(Continued)

Table 8.1:
(Continued)

Q1 Age	Age	Frequency	Percent	Valid percent	Cumulative percent
	42	8	3.3	3.3	81.1
	43	6	2.4	2.5	83.5
	44	6	2.4	2.5	86.0
	45	6	2.4	2.5	88.5
	46	3	1.2	1.2	89.7
	47	7	2.8	2.9	92.6
	48	6	2.4	2.5	95.1
	49	4	1.6	1.6	96.7
	50	2	.8	.8	97.5
	51	1	.4	.4	97.9
	52	1	.4	.4	98.4
	53	1	.4	.4	98.8
	57	2	.8	.8	99.6
	64	1	.4	.4	100.0
	Total	243	98.8	100.0	
Missing	System	3	1.2		
Total		246	100.0		

Table 8.2:
Summary statistics
of respondents'
ages

Statistics		
AGE		
N	Valid	243
	missing	3
Mean		34.66
Std. error of mean		0.540
Median		35.00
Std. deviation		8.412
Percentiles	25	28.00
	50	35.00
	75	41.00

Missing values

Not all your respondents will be willing to provide all the
information you ask for. Age can be a particularly sensitive

question and may be left out. This then becomes a *missing value*. SPSS automatically identifies missing values and lists them separately as you can see (3 cases). It will regard these cases as invalid and will exclude them from any analysis that requires this information.

On other occasions you may want to exclude cases that have a particular value and instruct SPSS to classify this value as a *missing value*. For example, you may wish to exclude everyone over 50 years of age from a particular analysis. If you now create a frequency table from the data, there will now be 9 missing values and only 237 valid cases in any subsequent analysis.

Cumulative percentage (Cumulative frequency)

The first three column headings are straightforward. *Frequency* is the number of cases of a given category of response; *Percentage* represents the *frequency* as a percentage of the total number of respondents. *Valid percentage* represents the *frequency* as a percentage of valid cases (i.e. it excludes the *missing values*).

Cumulative percentage adds up the *valid percentage* as it goes along, i.e. it adds the previous total percentage of valid cases to the current percentage. For example, 21-year olds formed 1.6% of valid cases, and 22-year olds formed 4.1%. Therefore, 5.8% of cases are 22 years of age or less. So, by the same process, 28.4% of cases are 28-years old or less; 77.8% of cases are 41-years old or less; and 100% of cases are 64-years old or less. *Cumulative percentage* is a useful measure for a number of reasons. For example, census returns to the **Office for National Statistics** are used to determine the age structure of the UK population, From these, insurance companies can construct **life tables** that show the *cumulative frequency* of age within the population (often separately for male and female, and even by geographical location). On the basis of these tables, insurance premiums can be calculated using the probability of surviving to or beyond a particular age!

Creating grouped frequency tables

The problem with **Table 8.1** is that it has so many categories, most of which have low numbers, that it is difficult to see any pattern. It is too diffuse. With some variables, the number of categories is often small so that this is not a problem. *Interval* variables often present this difficulty. One way around this is to group the data into a smaller number of categories.

For example, you might group together the ages into categories of 5 years as shown in **Table 8.3**. This table was produced using SPSS. Note that the end value of each group is the same as the first value in the next group up. This is because age is a continuum in that you are 25 on the day before your 26th birthday. In other words by stating groups in this way, it is understood that the first group goes from 21 to 25 years and 364 days!

▶
Table 8.3: Frequency table based on grouped data

Age Group		Frequency	Percent	Valid percent	Cumulative percent
Valid	21–25	43	17.5	17.7	17.7
	26–30	40	16.3	16.5	34.2
	31–35	49	19.9	20.2	54.3
	36–40	50	20.3	20.6	74.9
	41–45	33	13.4	13.6	88.5
	46–50	22	8.9	9.1	97.5
	51–55	3	1.2	1.2	98.8
	56–60	2	.8	.8	99.6
	61–66	1	.4	.4	100.0
	Total	**243**	**98.8**	**100.0**	
Missing	**System**	**3**	**1.2**		
Total			**246**	**100.0**	

The pattern of distribution is much clearer in **Table 8.3**. There is a trade-off. On the one hand you will gain a clearer picture of the distribution of the data, but in the process you will lose the detail. Where possible, in any analysis of *interval* data, it is best to use ungrouped data, but for presenting data to your readers, grouping it like this can make it easier for the readership to understand and see patterns.

As to deciding how to collapse the data into groups, there are no hard and fast rules. You need to be clear about what you are attempting to show with the data. However, you must also keep in mind that the greater the range within the group, the less detail can be seen. For example, if the groups had been based on ten years rather than five, there would have been far fewer categories and the nature of the data distribution would have been masked.

Graphical exploration of data

Histograms

The most well known form of chart that is described in this chapter is the *histogram*. Data is presented as a series of vertical bars, the length of which correspond to the frequencies. A *histogram* represents grouped data—in this case the groups are not the same as those used in **Table 8.2**. **Figure 8.1** is a *histogram* produced by SPSS from the data shown in **Table 8.1**.

Technically speaking, the area of each bar actually represents the frequency of respondents as a proportion of the entire sample. The total area of all the bars therefore represents the sample as a whole. As such, this is visually more effective at showing patterns of distribution than

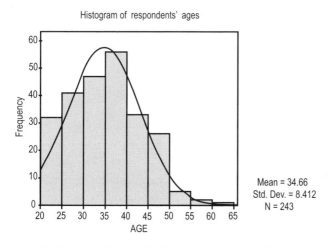

Figure 8.1: Histogram with *normal distribution curve* superimposed

tables of data. SPSS has a very useful facility of allowing you to put a *normal distribution* curve over the histogram so you can see if the distribution of your data approximates the *normal distribution* pattern. This distribution broadly approximates *normal distribution*, although because the data starts at age 21 so the distribution cannot continue down to zero while in theory if can continue much further to the right (e.g. there are some higher education students in their 70's and 80's).

Stem-and-Leaf chart

While most people will recognise a histogram, a *stem-and-leaf* chart is rather less well known. **Figure 8.2** is a *stem-and-leaf* chart based on the data in **Table 8.1**. *Stem-and-leaf* charts have the dual effect of representing the frequency of each category in a way that allows you to visually check the distribution and also the number of cases within each category of the variable concerned.

The *stem* column usually represents the first digit of the category, and in the case of fig 8.2 at the bottom of the chart you are told that the *stem* width is 10, so a *stem* of 2 really means 20 in this example. You are also told that each leaf represents one case. So, taking the first row where there is a frequency of 4, these are all 21 year olds. In the second row, of the 22 listed, 10 cases are 22 years old and 12 are 23 years old, and so on. The *stem* width is made as large as is compatible with the data. For example, if the chart was representing data on people's height, a height of 173 cm might be represented as:

Stem	Leaf
17	3

Stem width = 10
Leaf = 1 case

Figure 8.2 was generated by SPSS, which has also identified *extreme* values, or *outliers*. This is dealt with a little later in the chapter, but essentially they are values that lie some considerable distance away from the main distribution of the data and should be investigated.

AGE

```
AGE Stem-and-Leaf Plot

 Frequency    Stem &  Leaf

     4.00      2  .  1111
    22.00      2  .  2222222222333333333333
    17.00      2  .  44444455555555555
    16.00      2  .  6666677777777777
    14.00      2  .  88888888889999
    22.00      3  .  0000000000111111111111
    13.00      3  .  2222222333333
    24.00      3  .  444444444444555555555555
    28.00      3  .  6666666666666667777777777777
    16.00      3  .  8888888999999999
    13.00      4  .  0000001111111
    14.00      4  .  22222222333333
    12.00      4  .  444444555555
    10.00      4  .  6667777777
    10.00      4  .  8888889999
     3.00      5  .  001
     2.00      5  .  23
      .00      5  .
     2.00      5  .  77
     1.00  Extremes   (>=64)

 Stem width:      10
 Each leaf:        1 case(s)
```

Figure 8.2: Stem-and-leaf plot

Boxplots

Boxplots look at the distribution of data in a slightly different way. They chart the distribution on the basis of certain key measurements—essentially *central tendency* (which is discussed in the next chapter). These measurements are the *median, interquartile range,* and the *minimum* and *maximum* values. **Figure 8.3** is a *boxplot* of the data listed in **Table 8.1** and was generated by SPSS.

The shaded rectangular area is the *box* and the two vertical lines that extend upwards like a 'T' and downwards like an inverted 'T' are called *whiskers.* The bold horizontal line cutting across the box marks the position of the *median* value. The upper and lower edges of the box represent the third and first *quartiles* respectively. The *box* therefore represents the *interquartile range.*

The *whiskers* extend out to what the software has calculated to be the maximum and minimum values. Note that these values are not necessarily the same as the actual values, as

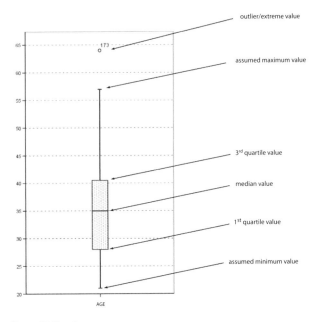

Figure 8.3: Boxplot

there is one respondent aged 64. As with the *stem-and-leaf* plot, this individual value has been identified as an *outlier* or *extreme value.* A very useful feature of the *boxplot* is that outliers are identified with the database entry number—i.e. case number. In effect, a *boxplot* warns you about using these cases for any meaningful calculations so that you can go back to the database to identify the original source of the data and check this value against the records (e.g. a specific questionnaire). This allows you to determine if a data entry error has been made or if there has been a mis-recording in the questionnaire.

> A very useful feature of the boxplot is that outliers are identified with the database entry number. This allows you to go back to the database and check for possible data entry errors with that record

Boxplots will exclude *missing values* when constructing the plot.

Cleaning up the data set

Frequency tables, *boxplots,* and *stem and leaf* charts provide you with valuable information about your data, particularly with respect to values that appear to be unreal or are at the extreme limits of the data. For example, if one of the cells in your database gave the age of a respondent as 993, all of these methods of summarising data would have picked it up. Common sense will tell you that this is a data entry mistake. Another example of this could be in a question that requires a 'yes' or 'no'. If the coding options are 0 = No, 1 = Yes, 9 = No response, but you find a code of 3 in the tables or charts, you would regard this as an error. Such errors are called *rogue values*—as would be an age of 993!

However, there will be other occasions when the data is somewhat removed from the main bulk of the data values, but you cannot reasonably say that it is an error. In **Table 8.1**, one mature undergraduate has given her age as 64. It is perfectly reasonable to suppose this is genuine record of a 64 year-old undergraduate (there are actually quite a number in higher education—and even older), so you cannot call it an error. A value that is reasonable, but which lies far removed from the main data is called an *outlier* or *extreme value*.

How a researcher deals with these values both in terms of classifying a case as a rogue or outlier value, and justifying the treatment of that data in any subsequent analysis, may be the subject of debate with colleagues—or criticism if there is disagreement about the treatment or any interpretation of data influenced by that decision. These decisions should be well-informed and justifiable on theoretical grounds.

Dealing with outliers & rogue values

Rogue values should be corrected whenever possible. This will involve you going back to the original questionnaire—if you are able to identify it—and check the appropriate response. The database record can then corrected. If this cannot be done, for whatever reason, it is not justifiable to guess what the response should have been unless there are very compelling reasons for being able to do so. This is widely regarded as unethical (faking!) and, anyway, it introduces bias into the process. It becomes your response rather than that of the respondent. In such circumstances, it is better to treat that response as a *missing value*.

131

Outliers can present something of a dilemma. In certain circumstances it may be best to regard *outliers* as *missing values* because of their potential to distort the findings of your survey. For example, if you are looking at the relationship between employment and participation in higher education, you may want to exclude the 64 year-old woman. On the other hand, if you are looking at general motivational factors, you may want to include her. *Outliers* can have a significant effect on the calculation of some key descriptive statistics—most notably the *mean* and consequently any tests of statistical significance based on the comparison of *means* (see chapter 10). In such cases, they should be excluded from the analysis, providing you give reasons why this is being done, and they can be justified on theoretical grounds.

Trim

Because the effects of *outliers* and *extreme values* can be quite marked, many researchers will apply a *trim* to the research data. This is much easier than seeking out and excluding individual records and SPSS and MINITAB include a facility to remove the highest and lowest 5% of values from any calculation of the mean. This produces something called the *trimmed mean*. However, you should only apply this where it is clear there are extreme values that may in same way distort the analysis of the data. Using the exploratory techniques described in this chapter will enable you to determine the appropriateness of this treatment.

SPSS tutorial

Charts

1. SPSS will generate a table of descriptive statistics, histogram, stem and leaf and boxplot charts from **Analyze > Descriptive statistics > Explore** (Fig 8.4)
2. Select the variable to explore, click the arrow (labelled) to transfer the variable to the Dependant List box. Click **Plots** to specify chart options (fig 8.5).

3. In the **Explore:Plots** dialog box make sure the checkboxes in the circled area of Fig 8.6 are selected, click **Continue**, taking you back to the **Explore** dialog box. Click OK to view the results.

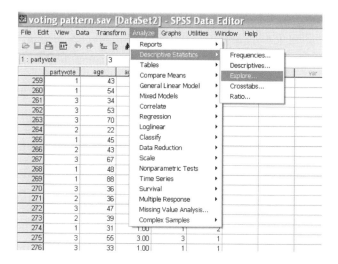

Figure 8.4: Analyze > Descriptive Statistics > Explore...

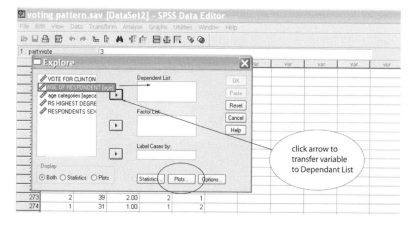

Figure 8.5: Select plots to open the graphs dialog box

4. The SPSS output window contains the various plots and summary descriptive statistics (fig 8.7). You can save the output window so that it can be viewed or printed later.

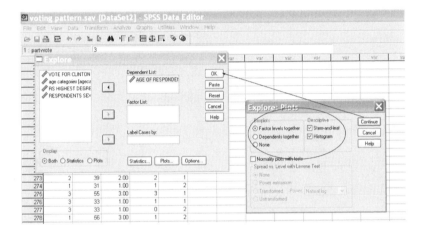

Figure 8.6: Selecting the type of plots required (stem-and-leaf and histogram in this case)

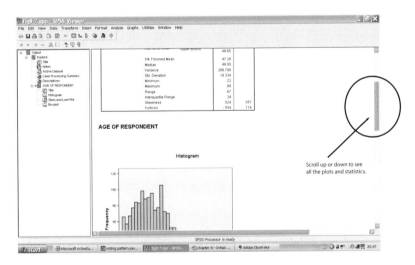

Figure 8.7: The SPSS Output window showing all the tables and plots asked for

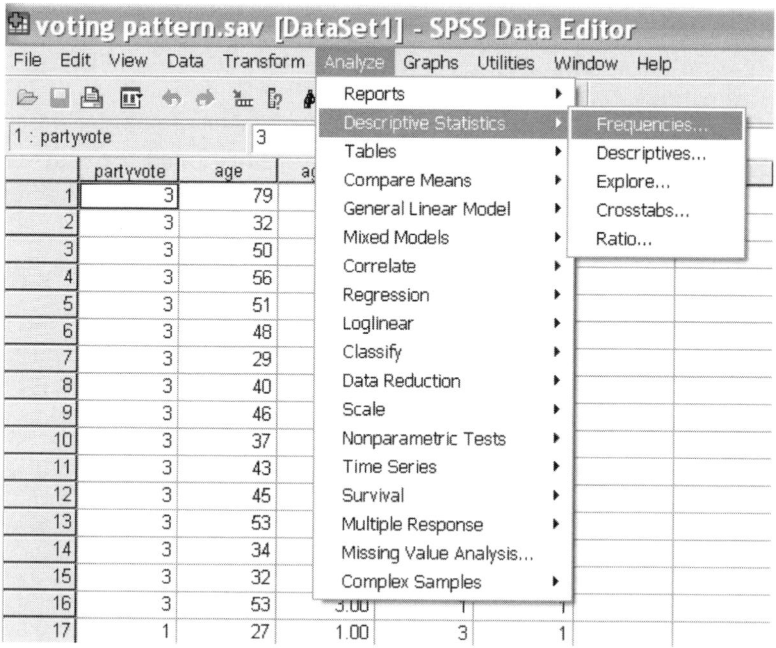

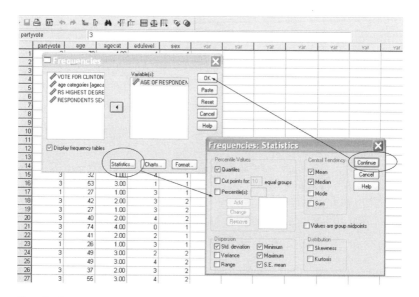

Figure 8.8: Analyze > Descriptive Statistics > Frequencies

Figure 8.9: Specifying the statistics required in addition to the frequencies table

Frequency table

5. A frequency table is created from the **Analyze >
   Frequencies** menu option (fig 8.8).

6. Transfer the variable to be explored in the *Variables* box,
   click *Statistics* to select the summary statistics required
   before returning to the **Frequencies** dialog box. Clicking
   OK will generate the table (fig 8.9 and 8.10). In this case,
   the statistics required are: mean, median, quartiles and
   standard deviation.

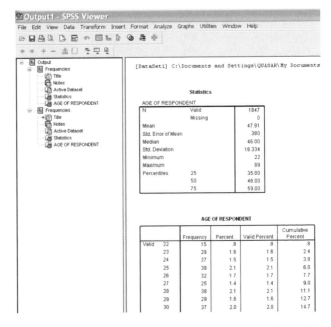

Figure 8.10: The SPSS Output window showing the statistics specified and the frequency table

Exercises

Progress questions

1. What are meant by the terms: *frequency, missing values, grouped data, cumulative frequency*?
2. What is meant by the terms: *outlier,* and *rogue value*?
3. What factors will you take into account when considering?
4. What are the advantages of displaying data in each of the following forms: *frequency table, histogram, stem and leaf chart, boxplot*?

Discussion points

1. In a frequency table where the variable values range from 1 to 100, it is cumbersome to try and present the frequency of each value in a report. You need to group the data. Suggest possible groupings, with a justification for your choices, if the variables were (a) annual number of visits to the cinema; (b) weekly shopping bill in pounds sterling (c) number of cars coming off the production line every day. **CLUE:** This discussion will revolve around whether or not a variable is a continuum as this will determine cut-off and starting points for each data group.
2. You have to produce a short report comparing the salaries of staff in different organisations and to present the data in graphical format. Explain your choice of graphical representation, and include an explanation of why you had rejected other forms of graphical representation.

Assignment

In chapter 2 you were asked to download a data file from the Studymates website (satisfaction.xls) and save it in SPSS. Now reopen the file and create frequency tables and exploratory charts of the interval data variables in the database. Comment on the following properties of the data:

a. Distribution in terms of closeness of fit to normal distribution (from the shape of the charts);
b. Distribution of the data as indicated by the mean, median, standard deviation and standard error.

9

Hypothesis testing—do your results show statistical significance?

Learning outcomes:

The aim of this chapter is to draw together a number of the ideas introduced in earlier ones so that the final two chapters can look in detail at the use of *inferential statistics*. The purpose of *inferential statistics* is to try and establish or test the likelihood of causal relationships between variables. By the end of this chapter you should:

- Understand the difference between causal and associative relationships between variables, including: spurious, intervening and interaction
- Understand the use of the 'null-hypothesis' versus 'research hypothesis' in hypothesis testing;
- Understand Type I and Type II errors and how they may be minimised;
- Understand the term 'statistical significance' and relate this to the probability of any given set of results occurring (i.e. the likelihood of any set of results);
- Understand the principles of 'significance levels' and how these are selected for any given purpose;
- Understand the role of 'one-tailed' and 'two-tailed' tests.

Hypothesis testing

The final two chapters will explain the tools available to the social researcher for testing *research hypotheses*. After all, the whole point of designing quantitative research is to test your ideas about the relationship between factors, After collecting the data, you need to interrogate it to look for relationships between variables that would support your *hypothesis*. This is called *hypothesis testing*.

However, it is not quite as simple as this and it is very easy to misread the nature of the relationship between variables and this often leads to very fundamental error, which is to draw false conclusions. It is important to understand the different kinds of relationships between variables.

Causal and associative relationships

There are several possible ways of interpreting, or explaining, an observed relationship between variables.

Spurious relationship

An observed relationship between two variables, 'A' and 'B', may actually be caused by another variable, 'C', that has not been observed or whose effect has not been recognised. There is, in fact, no causal relationship between the observed variables 'A' and 'B'. **Figure 9.1** illustrates this. Perhaps the easiest way to explain what can sometimes be an embarrassing error is to use an obvious and trivial example.

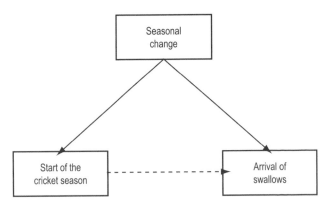

Figure 9.1: Spurious causal relationship between the start of the cricket season and the arrival of migrating swallows

There is an apparent association between the duration of the cricket season and the migratory behaviour of swallows. A fundamental error would be to assume that one caused the other! The true causal factor is the changing seasons. The message is simple. Treat all observed relationships with caution until you have eliminated, as best you can, all other possible causal relationships.

In practical terms, this kind of error is so easy to make as it is always possible that the researcher has not identified and collected data on the real causal variable—failed to notice that there is another factor at play here.

Intervening variables

There may be a causal relationship between two or more variables, but it is not necessarily a direct one. There may be an *intervening variable*. **Figure 9.2** illustrates what happens in this case.

One observed relationship frequently reported is between gender and career progression. It is often noted that women do not in always progress as far as men up their respective career ladders. One of the stated reasons for this observation is the role that women play within the family, so family commitments are an *intervening variable* that has a causal impact on career progression. While there is a direct association between gender and careers, this is as a consequence of one or more *intervening variables*. Since it is women that usually take on that family role, the effect is more usually observed with them. It is probable that if a male took on the role described by the intervening variable (and an increasing number of men have been doing so), you would observe a similar effect on their career progression.

Figure 9.2: The effect of an intervening variable

Interacting variables

It is often very difficult to pinpoint one causal association because human physical and psychological make-up, social circumstances and opportunities have profound effects on the way an individual views, interprets and reacts to the world. This means that there are frequently several or many variables that *interact* with each other and influence the observed relationship between two variables. Chapters 10 and 11 will explore this effect and how you can investigate the relationships between multiple variables.

For example, looking back at **figure 9.2**, there are probably influences other than the fact that an individual has childcare responsibilities. The opportunity to work and bring up children at the same time, the attitudes of partners towards work and parental responsibilities, the financial status of the family group, and perhaps many other factors will interact and influence the observed relationship.

An important function of social research is to identify the influences and to explain them. Statistics cannot explain, but they can go a long way towards identifying potential causal relationships.

The null hypothesis v. the research hypothesis

If your *research hypothesis* is **job satisfaction increases as employees' salaries increase**, you will want to look for evidence in the research data. For example, if you have devised a **job satisfaction index** you would compare this with **employee salaries** to see if there is a direct relationship—that is, the job satisfaction index will be highest in respondents whose salaries are highest.

Statistical significance

The problem is that while you may observe a relationship, this is not proof that one actually exists. *Sampling error* may have led to this result through chance, because your data has been collected from a sample, not the entire population. To get around this problem, you have to return to probability theory to calculate the **likelihood** of your results being obtained by chance.

The test invoked is called *statistical significance* and it involves comparing the results of your *research hypothesis* with the results of the *null hypothesis*. The null hypothesis asserts that there is **no relationship** between the variables. The test of *significance* is to calculate the probability of obtaining your results by chance. If the probability of having gained your observations by chance is high, you accept the *null hypothesis* because of the level of uncertainty. In other words, you should reject your *research hypothesis.*

By convention, it is accepted that you would only **tentatively** accept the *research hypothesis* and reject the *null hypothesis* if the probability of obtaining your results by chance alone was less that 5% (p < 0.05). In other words, you would be at least 95% confident that your research hypothesis is valid—this is referred to as *95% confidence level.* There are two other levels frequently used: 1% (p < 0.01, 99% confidence level) and 0.1% (p < 0.001, 99.9% confidence level). The lower the probability that your results were obtained through chance, the happier you should be with your hypothesis. The risk still remains that you can accept or reject your research hypothesis in error, but in good faith on the basis of *statistical significance* and you must always be cautious about your decisions. There is always an element of uncertainty.

In practice, the confidence level is chosen depending on the potential consequences of making a mistake. If accepting the *research hypothesis* would lead to radical changes in social policy affecting the lives of very many people, you would want to exercise caution and use p < 0.01 (1%) or even p < 0.001 (0.1%) as your threshold. You would want to reduce the risk of committing an error the consequences of which could be very unfortunate.

Type I and Type II errors

Even after assessing *statistical significance* there is still the risk of accepting or rejecting the *research hypothesis* in error. Statisticians have identified two types of error.

A *Type I* (type one) error occurs if the *null hypothesis* is rejected when it is in fact correct. This is also sometimes referred to as a *false positive* since the researcher has accepted a false *research hypothesis*. A *Type II* (type two) error occurs if the *null hypothesis* is accepted when in fact the *research hypothesis* is correct. This is sometimes referred to as a *false negative* as the researcher has rejected a correct *research hypothesis*. **Table 9.1** summarises this principle.

▶

Table 9.1:
Type I and
Type II errors

		In Reality	
		Null Hypothesis correct	Null Hypothesis incorrect
Research Observations	Null Hypothesis Rejected	**Type I error**	Correct decision
	Null Hypothesis accepted	Correct decision	**Type II error**

Keeping in mind that the researcher is testing the research hypothesis against the null hypothesis, the research hypothesis should only be tentatively accepted if the probability of the results supporting the hypothesis by chance is low—i.e. 5%, 1% or 0.1%. The *confidence level* is set in accordance with the level of risk of committing a *Type I* error. While the 95% confidence level is the normal threshold, if the consequences of committing a *Type I error* are potentially severe (e.g. in the field of social policy), the researcher may wish to set the acceptable confidence level at 99%.

One and two-tailed test of significance

Consider these two hypotheses:

Hypothesis 1: Driver reaction times become increasingly slower with increasing amounts of sleep deprivation.

Hypothesis 2: Driver reaction times are affected by sleep deprivation.

Hypothesis 1 is a *directional hypothesis*. That means that the hypothesis is predicting the direction of the relationship between the two variables. Reaction time is the *dependent variable* because the results measured are dependent on the amount of sleep deprivation, the *independent variable*. More on this will be covered in the next two chapters, but in principle the idea will be to control changes in the sleep deprivation pattern of a sample of people and then measure their reaction times. **Figure 9.3**, is the theoretical normal distribution curve for all possible observations made in that experiment. Results falling in the *region of rejection* (i.e. in the highest 5% of possible observations demonstrate that there is a 95% probability that this is not the result of chance. We would tentatively accept the research hypothesis that reaction times increase with increasing amounts of sleep deprivation and **reject** the null hypothesis. The onus on the researcher is to justify the rejection of the null hypothesis.

In the case of **figure 9.3**, the *research hypothesis* is predicting that the result will be at the higher end of the distribution. Equally, your hypothesis might predict an outcome at the lower end of the distribution. For example, you may have a hypothesis that predicts absenteeism from work is lower in those employees who are highly motivated.

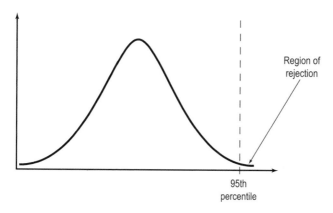

Figure 9.3: Region of rejection—directional hypothesis

In this case the *region of rejection* will be any value lower than that of the 5th percentile.

Hypothesis 2 is a *non-directional hypothesis.* It predicts a link between the two variables but does not indicate the directional nature of the link. According to this hypothesis a **reduction** in reaction times is as valid as an **increase**. In other words the researcher is saying that (s)he thinks there is a relationship, but not sure of what it is! This is a weak hypothesis and so the 5% of extreme values used to determine the rejection of the null hypothesis has to be shared between both tails of the distribution (see **fig 9.4**).

The regions of rejection are the bottom and top 2.5% of possible observations. In both examples, the risk of committing a Type I error is 5% or less. In the last two chapters, you will be exploring specific statistical tests of significance in more detail and they all work on the basis of identifying where observations fall within the *region of rejection* and work for both directional and non-directional hypotheses. However, if you are working with a *directional hypothesis* you will specify the test to apply a *one-tailed test* (i.e. because the observations will fall in a region of rejection that is at one end of the theoretical distribution. For *non-directional hypotheses* you will specify a *two-tailed test* (i.e. the region of rejection will be at both ends of the theoretical distribution of observations).

We will come back to this in the final two chapters.

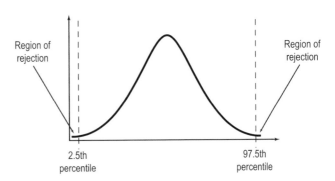

Figure 9.4: Regions of rejection—non-directional hypothesis

Degrees of freedom

Degrees of freedom refer to the number of components in our results that are free to vary. It is a difficult concept to master; however, a simple example will illustrate the principle.

Table 9.2 shows the results of a hypothetical survey into the colours of cars preferred by men.

Preferred Colour	Frequency
Black	25
Blue	61
Green	15
Red	85
White	40
Yellow	5
Total	**231**

◀ Table 9.2: Colour preferences

You want to know if this distribution of choices is statistically significant and shows that men are likely to be attracted to buying red cars in preference to other colours. To determine this you need to know the *degrees of freedom* and then apply your statistical test. The result of a test is a number that you need to look up against the *degrees of freedom* in tables designed for the test. There are *five degrees of freedom* in this example because if the value of one of the categories was known, the other **five** are free to vary so long as the total came to 231.

The 'rule of thumb' approach is to calculate the *degrees of freedom* as:

Degrees of freedom = number of categories – 1

Parametric and non-parametric tests

There are several common tests used to assess the statistical significance of your findings. The choice of test is determined by the nature of the data you are working with. Many text books will give the formulae required for these tests, but the use of computer software packages make the task quicker and arguably with less risk of error than working by hand. It is more important to understand the principles and to be able to apply the appropriate test to any given set of circumstances.

There are two major groups of tests: *parametric* and *non-parametric*. Tests of significance do assume that you have used random sampling techniques to draw your research sample.

Parametric tests

These tests are applied when you can make two assumptions about your data:

1. the samples have been drawn from a population in which the variables concerned are normally distributed or are very nearly so;
2. the variables are on an *interval* scale.

Non-parametric tests

These tests make much weaker assumptions about the data. The data does not have to be *interval* or be normally distributed.

Exercises

Questions

1. When exploring causal relationships, how do the terms *spurious, intervening* and *interacting* variables explain possible errors?
2. What is meant by the term *test of statistical significance*?
3. What are the main assumptions that underpin (a) *parametric* and (b)*non-parametric* tests of significance?
4. Under what circumstances would you use a (a) *one-tailed test* (b) *two-tailed test of significance*?

Discussion point

In an attempt to minimise the risk of committing a Type I error, a researcher decides to use $p < 0.01$ rather than $p < 0.05$ as the test for rejecting the *null hypothesis*. By doing so, though, the researcher increases the risk of committing a Type II error. Discuss why that is the case. In what circumstances do you think a Type II error is preferable to a Type I error?

10

Exploring the relationship between two variables

Learning outcomes:

The whole point of using statistical analysis in the social sciences is to determine or test a relationship between two or more factors. This chapter focuses on assessing the nature of the relationship between two variables—and this is referred to as *bivariate analysis*. The aim of this chapter is not to tackle the mathematics of the tests (this is quite a difficult area) but to enable you to understand why we apply tests, and which tests to apply depending upon the research design and data analysis needs.

All the datasets used in this chapter are available from Studymates.

By the end of this chapter you should:

- Understand that bivariate analysis explores the relationship between two variables (dependent and independent);
- Understand when to use 'parametric' and 'non-parametric' tests of statistical significance in bivariate analysis;
- Know and be able to apply common 'parametric' and 'non-parametric' tests of 'statistical significance';
- Be able to use SPSS to explore bivariate relationships.

The relationship between two variables

Chapter 9 looked at the idea of *statistical significance* as a way of evaluating the likelihood of a real relationship between two variables rather than a result produced by chance. While you can never be absolutely certain, the stronger the *statistical significance*, the greater the confidence you can have in rejecting the *null hypothesis*. It is not sufficient to produce data that shows one as one variable varies there is variation in the other. You have to show that there is a low probability that the data showing a clear relationship is the result of chance.

Consider **Table 10.1** which summarizes a hypothetical sample of voters' intentions in an election.

			VOTE FOR PARTY A, B, or C			
			Party A	**Party B**	**Party C**	**Total**
age categories	18–34	Count	153	99	186	438
		% within age categories	34.9%	22.6%	42.5%	100.0%
	35–44	Count	156	81	207	444
		% within age categories	35.1%	18.2%	46.6%	100.0%
	45–64	Count	219	82	316	617
		% within age categories	35.5%	13.3%	51.2%	100.0%
	65+	Count	133	16	199	348
		% within age categories	38.2%	4.6%	57.2%	100.0%
Total		Count	661	278	908	1847
		% within age categories	35.8%	15.1%	49.2%	100.0%

▶ Table 10.1: Hypothetical voting intentions by age groups based on a ample of voters

The age categories are grouped data—note that they are not equal sizes—and the rows show the proportion of the total number of people in each age group who voted for a particular party. In other words, the table is showing the apparent voting preferences within each age group as a percentage of that group (the actual numbers are also shown), The problem with this table is that while there are some very interesting patterns such as:

- Party B is the least popular Party.
- Older people, and especially those over the age of 65, have a particular dislike of Party B.

- All age groups seem to have a preference for Party C (which has taken 49.2% of the total vote in this sample).

Now, this is all very interesting **but** these data are derived from a sample and we cannot tell if the data shows any *statistical significance*. Unless this information is provided, nothing can be inferred from the data regarding the possible outcome of the election.

One or more tests of significance must be applied and published with the data.

Testing for statistical significance

There are a number of tests that can be applied to research data based on bivariate analysis and this chapter will describe and explain the use the most commonly used methods in social science research. The selection of a test depends on the nature of the data being explored. At the end of chapter 9, the terms *parametric* and *non-parametric tests* were used and a brief overview of the conditions under which each type would be applied was given. To recap, these were:

- *Parametric tests*: The data must be *interval* and *normally distributed* or nearly so,
- *Non-parametric*: The data is *ordinal* or *nominal* or *interval data* that does not conform to or approach *normal distribution*.

The first decision you have to make is which of these types to apply. This is an area of considerable debate among statisticians and is not clear-cut, and some statisticians will use *parametric* tests with ordinal data. This becomes more sensible if there are a large number of categories—e.g. a scale of 10—or where ordinal data is aggregated to form a scale. The **satisfaction** database used in earlier chapters— and is available from Studymates, has a variable called *satisfaction* that is derived by aggregating the responses from a number of other Likert scale variables. This results in a range of values across the sample that are effectively interval in nature.

Another factor to take into account is the nature of the groups you are comparing—and is linked to the way you have designed the research:

- *Independent samples*: The research is based on collecting data from two different groups of people. For example, you may want to compare the ages of students in two universities. The samples will be drawn from two separate populations—*independent samples*. This method is also referred to as *between-subjects* design.

- *Related samples*: Some research is based on collecting data from the same group of people at different times. For example, you may want to test the hypothesis that people's performance in a given activity is reduced by the effect of alcohol consumption. Clearly, you will be comparing data collected from the same group of people on two separate occasions. This approach is also referred to as *within-subjects* design.

While most research aims to compare two or more groups (e.g. gender, ethnicity, occupations), some research will only collect data from one group and this adds an extra dimension to the decision about which tests of significance to use.

Table 10.2 maps some of the main tests against the nature of the *dependent variable* and the number of groups or populations being compared.

Table 10.2: Mapping of tests against data type and number of comparison groups and research design

Dependent variable	Basis of tests	Comparison groups		
		1 group	Between—subjects design (Independent samples)	Within-subjects design (Related/paired samples)
Nominal data (non-parametric)	Tests based on comparing observed distributions against expected distributions if the null hypothesis were true	Chi-square or binomial test	Chi-square	Chi-square

(Continued)

Dependent variable	Basis of tests	Comparison groups		
		1 group	**Between—subjects design (Independent samples)**	**Within-subjects design (Related/paired samples)**
Ordinal data (non-parametric	Tests are based on ranking the data and comparing observed ranking against the expected ranking if the null hypothesis were true.	Chi-square	Mann-Whitney U	Wilcoxon Matched-Pairs Signed Ranks Test
Interval data (parametric)	Tests are based on comparing means and/or variances.	t-test	t-test or F-test	t-test or F-test

◄ **Table 10.2: (Continued)**

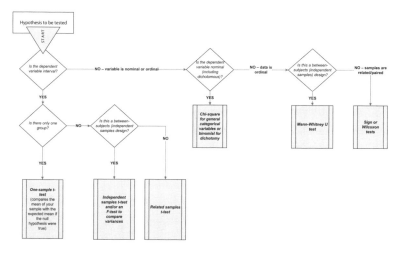

Figure 10.1: Flowchart to select appropriate tests of statistical significance

Table 10.2 lists the more commonly used tests of statistical significance in the social sciences. They are by no means the only ones, as statistics theory is constantly being debated and developed like any branch of science and mathematics. As the aim of this book is to equip you with an understanding of the practicalities of statistical tests, developing a working understanding of relevant statistical theory, a description of all the possible tests would be counter-productive to this understanding. The tests listed in Table 10.2 will be sufficient for almost all but the most complex of research projects.

Fig 10.1 is a decision-making flowchart that will help you identify the appropriate test, and the rest of the chapter will describe the tests in more detail and how they are applied (especially using SPSS).

Parametric tests

Independent (between-subjects) samples t-test

Research involving the analysis of *interval* data variables is generally concerned with looking at differences in two particular areas: means and variance (i.e. distribution of values). Essentially, we want to be able to identify statistical significance of the:

- Differences in mean values of the comparison groups; and/or
- Differences in the distribution (i.e. *variance*) of the data in the comparison groups.

In chapter 6, we discussed the hypothetical data on the ages of mature students in two colleges (College X and College Y). This is a typical example of an *independent or between-subjects* design—i.e. the two comparison groups are unrelated and were drawn randomly from separate *sampling frames*. The table (Table 10.3) is reproduced again below, with an added row showing the *variance*. *Variance* is calculated as the square of the *standard deviation* (see chapter 6).

▶ Table 10.3: Comparison of student populations in two colleges

	College X	College Y
Mean	34.7	35.0
Median	35.0	34.8
Standard Deviation (sd)	*8.4*	*4.3*
Variance (=sd²)	*70.6*	*18.5*

There seems to be little between them in terms of *mean* and *median* values: they are very similar. However, the difference in *standard deviation* is very marked and so each sample is showing different patterns of dispersion—but is this difference statistically significant?

The key test of statistical significance is the *t-test* which compares the means of the comparison groups and determines the probability of the difference arising as a result of chance—i.e. there is no significant difference between the groups. At the same time SPSS checks the *variances* in the data of the comparison groups and determines if they are equal—this is called the *F-test* and is a test for *equality of variance*, i.e. homogeneity. If this figure is statistically significant (i.e. $p < 0.05$) then the inference is that there is a real difference in the variance between the two groups—i.e the difference in standard deviation, for example, represents a real difference between the two groups and is not a random event. The *t-test* therefore provides you with two key pieces of data relating to statistical significant differences:

- Are the means of the two groups statistically significantly different (the *t* statistic)?
- Is any observed difference in variance statistically significant (the *F* statistic).

Look at this from a visual perspective. Fig 10.2 shows how the distributions of ages in both samples may look like if normal distribution is assumed.

College X seems to have a wider distribution of ages in its population compared with College Y.

The *t-statistic* is related to the *Z-score* described in chapter 6. A data value can be converted to a number of standard deviations above or below the population mean value. A Z-score of 1.65 above or below the mean places it is at the 95th or 5th percentile, respectively. See how this also relates to $p < 0.05$. *t* replaces Z in the *t-test* and represents the number of standard errors above or below the *standard error of differences between means*—i.e. the standard deviation from the true mean if the null hypothesis is correct . In effect, the test is being used to determine if the

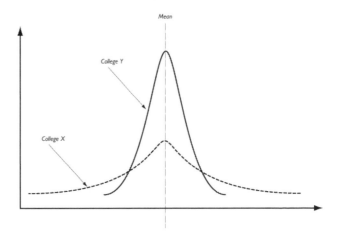

Figure 10.2: Theoretical distribution of student ages in College X and College Y

two samples have been drawn from identical populations or do represent a real difference.

The *standard error of differences between means* is dependent on the variances of the samples. Different results for t will be gained for samples with the same variance than those with different variances. The *F-statistic* tells us if the two samples have equal variance and so will tell us which value of t to use.

As this is not meant to be a mathematically focused book, put the technical reasons to one side because the hard work is now done by statistics software, and the interpretation of results and where they come from is more important than memorizing and using mathematical formulae.

Table 10.4 shows the times taken by 30 male and 30 female workers.

Table 10.4: Times taken to do a task by 30 male and 30 female workers	TIMES IN MINUTES TAKEN BY MALE WORKERS	TIMES IN MINUTES TAKEN BY FEMALE WORKERS
	18	14
	19	12
	14	10
(Continued)	16	12

TIMES IN MINUTES TAKEN BY MALE WORKERS	TIMES IN MINUTES TAKEN BY FEMALE WORKERS
12	8
18	10
16	10
18	8
16	12
19	16
16	14
16	12
15	15
25	18
10	12
15	14
18	22
12	12
10	10
16	12
17	16
18	19
21	15
9	9
14	12
13	13
20	23
11	12
16	18
19	23
Mean: 15.9	Mean: 13.8

◄ Table 10.4:
(Continued)

A visual inspection of the data shows that there is a difference in the mean time taken by male and female workers, with female workers taking less time on average than male workers to complete the task, but is this difference statistically significant? The hypothesis to be tested is:

Hypothesis: The mean time taken to complete task A by female workers is less than that taken by male workers.

This is a *between-subjects* or *independent samples* design as we will be comparing two entirely separate groups of workers (i.e. male and female workers).

Testing the hypothesis with SPSS

1. Fig 10.3 shows the variable view in SPSS. Data is recorded against the variables **gender** and **time**, and gender is *nominal* (a dichotomy) and time is *interval.*

2. The *independent t-sample test* is set up via the menu options: **Analyze** > **Compare Means** > **Independent-Samples T-Test** (fig 10.4).

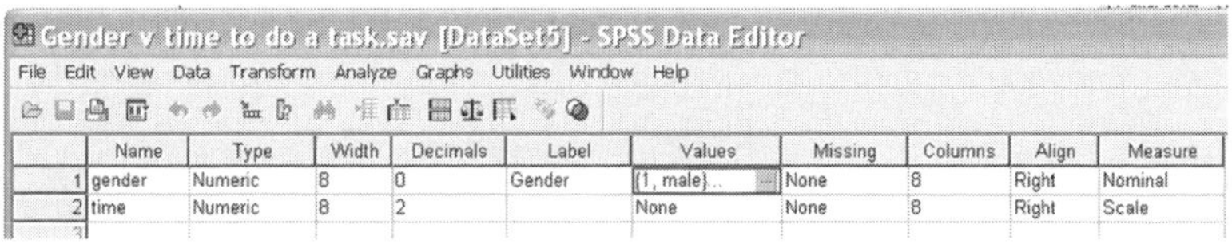

Figure 10.3: Setting the properties of a variable

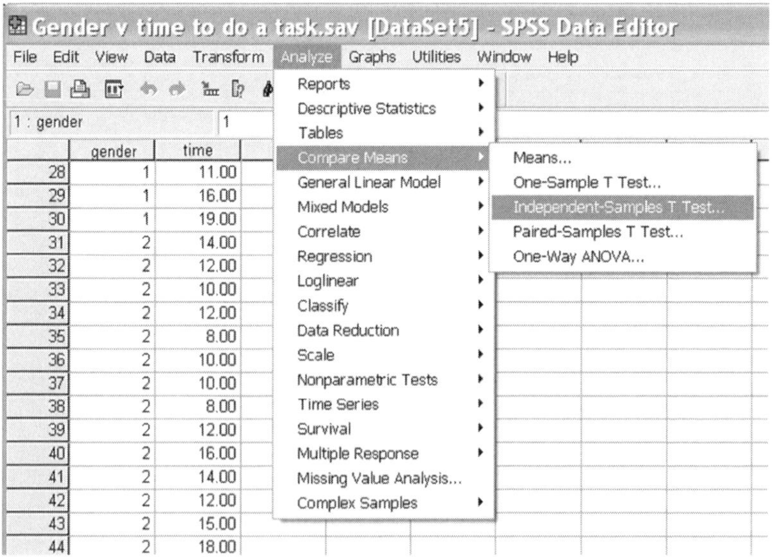

Figure 10.4: Menu route to independent-samples t-test

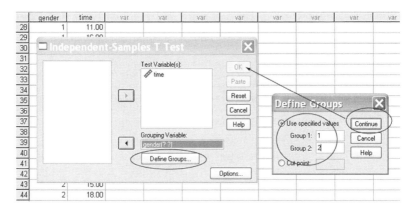

Figure 10.5: Defining the groups for the t-test

3. Click to select the variable **time** and transfer it from the left hand variable box to the **Test Variable(s)** (i.e. the dependent variable) box by clicking the arrow next to that box. Do the same to transfer the variable **gender** (the independent variable) to the **Grouping Variable** box. You will see (??) next to **gender** and this is because you have to define the groups. The independent variable is not always a dichotomy so the software needs to know which categories of the variable you want as comparison groups. Click **Continue** to return to the T-TEST box (fig 10.5), then click **OK** to run the test.

The SPSS printout (**Table 10.5**) gives two tables.

• **Group Statistics:** The basic characteristics of the two comparison groups showing that there is a difference in means, standard deviations and standard errors of the means.

• **Independent Samples Test:** This table will indicate if the difference in means is statistically significant and should be interpreted as follows.

Levene's Test for Equality of Variances produces the F-statistic referred to earlier, comparing the variances to see if they differ in a statistically significant way. This result shows there is a difference in variance ($F = .761$) but that it is not statistically significant (Sig $= .387$). *Therefore both groups are assumed to have equal variance. If F had shown significance at p<0.05, then we would conclude that the two groups were of unequal variance.*

t-test for Equality of Means now tells us if the difference in means is itself statistically significant. However, there are two rows of values, one where the variance of both groups are equal (the top row) and another where the variance of the groups is unequal. It has now been determined that the F-test shows the variances to be most probably equal, so the top row of figures is used. Had the F-statistic been significant at $p<0.05$, we would use the bottom row of figures for groups of unequal variance.

t has a value of 2.151 (i.e. effectively 2+ Z-scores above the mean value of the differences if the null hypothesis were true) and this makes the differences in means statistically significant at $p = 0.36$. However, the t-test assumes a two-tailed significance by default but the hypothesis being tested is a directional one. The mean time taken to complete a given task is significantly higher in males than females and there is a real difference in the work output of female workers as a result. For this we need the one-tailed probability and this is calculated simply by halving the two-tailed figure. In other words, the difference in means is actually statistically significant at $p = 0.018$ (i.e. still $p<0.05$, but more so than if we had taken a non-directional hypothesis).

The remaining figures in that row add more contextual information. The measured difference in the means is 2.13333 minutes (call it 2!) and the standard error of measurement (see chapter 6) of 0.99196 means that the real difference in means at the 95% confidence level lies between 0.15 and 4.12 minutes (rounding up).

Conclusion: We can reject the null hypothesis that male workers are not slower in carrying out a particular task in favour of accepting the hypothesis that female workers in the company are quicker at the task. The difference in mean times to complete the task is statistically significant at p<0.05.

You will see that some data in the results are reported to a level of accuracy that exceeds the ability to measure them! Therefore, in reporting these results you would do so at levels that are meaningful—not to 6 decimal places!—but also indicate the level of accuracy you have placed on the values.

Group statistics						
	Gender	N	Mean	Std. deviation	Std. error mean	
time	male	30	15.9000	3.55596	.64923	
	female	30	13.7667	4.10788	.74999	

Table 10.5: SPSS printout of independent t-test analysis

Independent samples test									
		Levene's test for equality of variances		t-test for equality of means					
		F	Sig.	t	df	Sig. (2-tailed)	Mean difference	Std. error difference	95% Confidence interval of the difference
									Lower	Upper
time	Equal variances assumed	.761	.387	2.151	58	0.36	2.13333	.99196	.14771	4.11896
	Equal variances not assumed			2.151	56.833	0.36	2.13333	.99196	.14684	4.11983

Paired (within-subjects) sample t-test

This is essentially a before and after scenario. For example, you may want to assess the effectiveness of a training programme on the length of time it takes to perform a task. **Table 10.6** shows the mean times taken to perform the task before and after training for a sample of 30 workers randomly selected from the trained workforce. The use of a t-test would be a very effective way of evaluating the training in some quantifiable way. If the differences in time are statistically significant, you could actually estimate the cost savings and productivity benefits resulting from the training.

WORKER	MEAN TIME IN SECONDS BEFORE TRAINING	MEAN TIME IN SECONDS AFTER TRAINING
1	18	14
2	19	12
3	14	10
4	16	12
5	12	8
6	18	10
7	16	10
8	18	8
9	16	12
10	19	16
11	16	14
12	16	12
13	15	15
14	25	18
15	10	12
16	15	14
17	18	22
18	12	12
19	10	10
20	16	12
21	17	16
22	18	19
23	21	15
24	9	9
25	14	12
26	13	13
27	20	23
28	11	12
29	16	18
30	19	23

In most cases, training has reduced the time taken to perform a task, but for others time has increased or remains static. This appears pretty conclusive in that the training programme does appear to have increased the speed of carrying out the task, but this is a sample of workers and the data may be the result of chance or sampling error. The hypothesis being tested is:

> Hypothesis: The impact of training has been to reduce the time it takes for workers to complete task A.

Testing the hypothesis with SPSS

1. Begin the analysis from **Analyze** > **Compare Means** > **Paired-Samples T Test**. In the Paired-Samples T Test dialog box, select the two **MEAN TIME ...** variables (their names will then appear in the Current Selections box to show you intend to compare the means of these variables (fig 10.6). Click the arrow to transfer them to the Paired Variables box and click **OK** to run the test.

2. The results of the analysis (Table 10.7) include the basic descriptive statistics in the Paired Samples Statistics. The second table, Paired Samples Correlations represents the

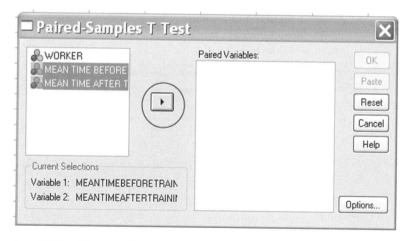

Figure 10.6: Specifying the variables for the paired-samples t-test

strength of the relationship between the two variables—but more will be said about correlations later in the chapter. The third table, Paired Samples Test, shows that the difference in the means is statistically significant. The default test is two-tailed, but again the hypothesis being tested is directional so the one-tailed significance is $p = 0.0015$ (i.e. $p<0.01$) and so is highly significant.

Conclusion: The null hypothesis that there was no improvement in the time to complete a given task following training is rejected in favour of the research hypothesis that completion times were significantly improved following training.

▶
Table 10.7:
SPSS printout
of the paired-
sampled t-test

Paired samples statistics		Mean	N	Std. deviation	Std. error mean
Pair 1	MEAN TIME BEFORE TRAINING	15.90	30	3.556	.649
	MEAN TIME AFTER TRAINING	13.77	30	4.108	.750

Paired samples correlations		N	Correlation	Sig.
Pair 1	MEAN TIME BEFORE TRAINING & MEAN TIME AFTER TRAINING	30	.563	.001

Paired samples test	Paired differences					t	df	Sig. (2-tailed)
		Mean	Std. deviation	Std. error mean	95% Confidence interval of the difference			
					Lower	Upper			
Pair 1	MEAN TIME BEFORE TRAINING - MEAN TIME AFTER TRAINING	2.133	3.617	.660	.783	3.484	3.230	29	.003

One-sample t-test

The aim of a one-sample t-test is to determine if your sample drawn from the population is representative of the general population. The approach is to compare the mean value of the test variable in your sample with the known mean of the population. For example, you may want to test if the claims by a company that they pay their workers above the average for their sector are justified. One way of testing this is to compare the mean value of salaries in the company with the known mean value for the sector[1]. For the purpose of this example, let us take the mean salary for the sector as £20,000.

This is an example of a situation where a small number of very highly paid or very low paid members of staff may distort the analysis. It is worth exploring extreme values and perhaps excluding them from the analysis (but you must explain why you are doing this).

Hypothesis: The mean salary of Company A's employees is significantly higher than the mean of the sector as a whole.

Testing the hypothesis in SPSS

1. Use **Analyze > Descriptive Statistics > Explore** (fig 10.7). Move **gender** to the Dependent variable, click Statistics (circled) and make sure the Outliers checkbox is ticked. Click Continue (circled) followed by OK in the Explore dialog box to run the analysis.

2. The results of the analysis will now include a table of extreme values (Table 10.8). In this example, it is the top 5 values that may distort the analysis. These are clearly the salaries of the top executives in the company and it would be best to exclude them from the t-test analysis. The best way to do this is to classify the highest values as *missing values* so the program ignores them.

---

[1] This is an over-simplification as there are a number of other things to consider—e.g. balance of job roles etc, but the idea is to demonstrate the use of the one-sample t-test.

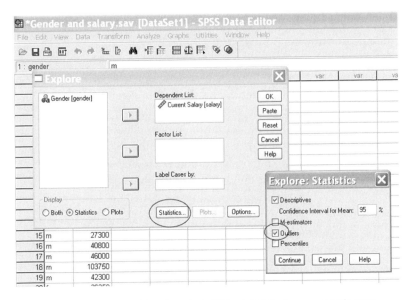

Figure 10.7: Specifying the variable and required additional statistics for a one-sample t-test

▶
Table 10.8:
Extreme values

Extreme values				
			Case number	Value
Current salary	Highest	1	29	135000
		2	32	110625
		3	18	103750
		4	343	103500
		5	446	100000
	Lowest	1	378	15750
		2	338	15900
		3	411	16200
		4	224	16200
		5	90	16200

3. It is best to create a new variable with the high outliers set as *missing values*. This way, the original data will not be destroyed. Use **Transform** > **Recode** > **Into Different Variables** ... In the dialog box that opens (fig 10.8), transfer

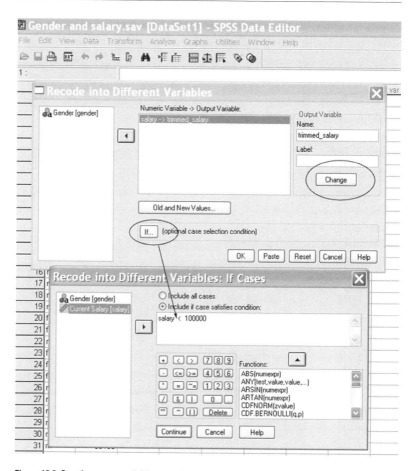

Figure 10.8: Creating a new variable to exclude extreme values (in this case very high salaries)

the **salary** variable to the Numeric variable -> Output
Variable box, give the new variable a name in the Output
Variable box and click Change. You now need to make
sure that salaries of £100 000 or higher are made missing
values. Click **If..**and in the new dialog box make sure the
Include if case satisfies condition and type ***salary < £100000***
in the box beneath it—this will ensure only salaries less
than £100 000 are included in the new variable. Click
Continue and then OK to run the recode instruction.

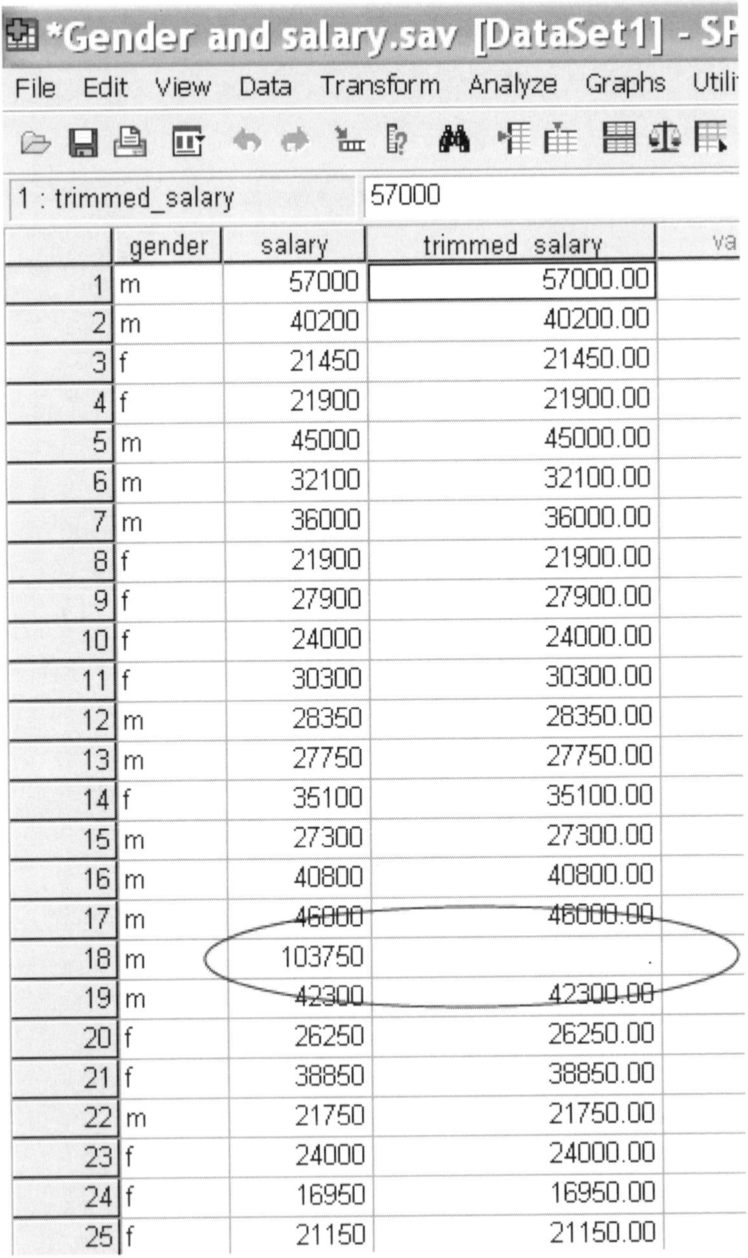

	gender	salary	trimmed_salary
1	m	57000	57000.00
2	m	40200	40200.00
3	f	21450	21450.00
4	f	21900	21900.00
5	m	45000	45000.00
6	m	32100	32100.00
7	m	36000	36000.00
8	f	21900	21900.00
9	f	27900	27900.00
10	f	24000	24000.00
11	f	30300	30300.00
12	m	28350	28350.00
13	m	27750	27750.00
14	f	35100	35100.00
15	m	27300	27300.00
16	m	40800	40800.00
17	m	46000	46000.00
18	m	103750	.
19	m	42300	42300.00
20	f	26250	26250.00
21	f	38850	38850.00
22	m	21750	21750.00
23	f	24000	24000.00
24	f	16950	16950.00
25	f	21150	21150.00

Figure 10.9: The new variable (*trimmed salary*) shows extreme values as *missing values* to exclude them from the calculations

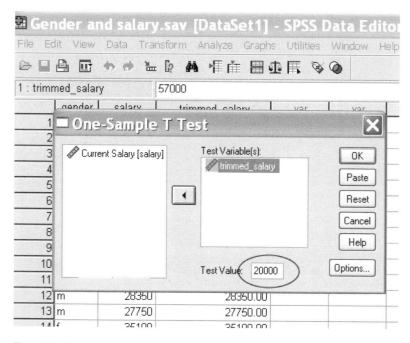

Figure 10.10: The one-sample t-test now uses the new variable and tests this against the known mean salary for the sector (i.e. £20 000)

4. The new variable is added and note that in fig 10.9 the high salaries are now missing values. This new variable is used in the one-sample t-test.

5. Perform a one-sample t-test from the **Analyze > Compare Means > One-Sample T TEST** menu. As shown in fig 10.10, transfer the new variable to the Test Variables box and type in the test value of £20 000 in the Test Value box (i.e. this is the known population mean salary for the sector) and click OK to run the test.

6. The result of the test (Table 10.9) shows that the trimmed mean salary in this company is £33 600, much higher than the £20 000 for the sector as a whole. In fact, the standard deviation of just under £15 200 is large and so around two-thirds of employees earn between £18 000 and £48 000 (approx). The t-test shows this to be highly significant (p<0.001).

One-sample statistics				
	N	Mean	Std. deviation	Std. error mean
trimmed_salary	469	33607.6759	15177.30299	700.82274

One-sample test						
Test value = 20000						
	t	df	Sig. (2-tailed)	Mean difference	95% Confidence interval of the difference	
					Lower	Upper
trimmed_ salary	19.417	468	.000	13607.67591	12230.5271	14984.8247

> Conclusion: We can reject the null hypothesis as the mean salary of employees in this company is very significantly higher than the sector as a whole (p<0.001).

Non-Parametric tests

Chi-Square (x^2)

This is a good general purpose non-parametric test for nominal data. You can use the *binomial test* for dichotomous data (see later). *Pearson's chi-square* is very widely used because it makes few assumptions about the nature of the data, but that makes it relatively weak when compared with the parametric tests described above.

The *chi-square (pronounced kie-square)* test involves comparing the observed frequencies of each category against the frequencies that you would have expected to observe if the *null hypothesis* were correct. The test generates the probability that the observed frequencies could have been obtained by chance—i.e. probability that the null hypothesis is correct.

SPSS will generate the chi-square statistic very easily, calculating the expected frequencies automatically, but you can also calculate it reasonably easily without the use of a computer program (which is why it is such a useful test).

Table 10.10 summarizes the results of a survey of 1620 responses to a question the researcher believes will show significant gender differences:

Q: Do you believe X would make a good leader of a political party?

	Male	Female	Totals
Yes	100	500	600
Don't know	50	70	120
No	150	750	900
Totals	300	1320	

◄ Table 10.10: Responses to a question on leadership quality

The raw numbers are quite difficult to work with because of the different gender group sizes, but there is a prima facia case for suggesting a gender bias. Note that the dependent variable is not a dichotomy—i.e. there are three possible responses: yes, no and don't know. To determine if there is a statistically significant difference in the views of male and female respondents, we need to know what the distribution of responses would be if the null hypothesis were true (i.e. no gender bias). This distribution is referred to as the *expected distribution or frequency* and we then compare the observed distribution with the expected.

Logically, if there is no difference in the views of males and females, then the distribution of responses of yes, no and don't know would be in the same ratio for each gender. The table of expected frequencies would look like Table 10.11. In other words, you redistribute the data in each column so that the ratio of frequencies in each column matches the ratios in the Totals column. In other words, the ratio of responses is the same for both groups—which would be expected if there was no difference between them. The *chi-square* test compares the observed data with the expected data and determines the probability of the difference being the result of chance—i.e. is it statistically significant.

Table 10.11:
Tables of
expected
frequencies
for the null
hypothesis

	Male	Female	Totals
Yes	111.1	488.9	600
Neutral	22.2	97.8	120
No	166.7	733.3	900
Totals	300	1320	

There is one word of warning, though. Small numbers present a problem with this test. If any of the cells of data have an expected value of less than 5, you need to use this test with caution. However, you can be less cautious if there are more than three categories involved in the test. The rule of thumb to adopt is that if 20% or more of the expected frequencies are less than 5, or any are less than 1, the chi-square test should not be used like this. You should regroup the data so that the expected frequencies of grouped categories satisfy these requirements. Clearly, you will need to modify your hypothesis so that it relates to the grouped categories rather than the original ones.

Hypothesis: The voting pattern in a recent election shows a significant effect of gender on voting preference.

Testing the hypothesis x^2 in SPSS

1. To compare voting pattern of the genders, a table comparing both genders is required. Create this from **Analyze > Descriptive Statistics > Crosstabs**. This will open a crosstabs dialog box (fig 10.11). The dependent variable (in this case **Vote for Party**) is transferred to the Row(s) box and the independent variable (in this case **Respondents' sex**) is transferred to the Column(s) box. Click Statistics to open the Statistics dialog box and make sure **Chi-square** is checked. Click Continue, then OK when returned to the Crosstabs dialog box to run the analysis.

2. The output from SPSS (Table 10.12) shows:
   a. A Case Processing Summary—simply information about valid and missing cases;
   b. A crosstabulation of the data showing observed voting numbers for each gender so there is a visual comparison;
   c. Chi-square tests results table. This shows the chi-square result to be statistically significant at $p<0.001$.

Conclusion: The null hypothesis may be rejected as the high statistical significance suggests that gender of the electorate has influenced the result of the election.

Case Processing Summary

	Cases					
	Valid		Missing		Total	
	N	Percent	N	Percent	N	Percent
VOTE FOR PARTY A, B, or C * RESPONDENTS SEX	1847	100.0%	0	.0%	1847	100.0%

◄ Table 10.12: Output form the Chi-square test in SPSS

VOTE FOR PARTY A, B, or C * RESPONDENTS SEX Crosstabulation

Count

		RESPONDENTS SEX		
		male	female	Total
VOTE FOR PARTY A, B, or C	Party A	315	346	661
	Party B	152	126	278
	Party C	337	571	908
Total		804	1043	1847

Chi-Square Tests

	Value	df	Asymp. Sig. (2-sided)
Pearson Chi-Square	33.830(a)	2	.000
Likelihood Ratio	33.866	2	.000

(Continued)

Table 10.12:
(Continued)

Chi-Square Tests			
	Value	df	Asymp. Sig. (2-sided)
Linear-by-Linear Association	19.360	1	.000
N of Valid Cases	1847		

ª *0 cells (.0%) have expected count less than 5. The minimum expected count is 121.01*

The binomial test

In chapter 3 probability was explored in detail and the example of spinning a coin was used. When it stops spinning, it will either land on heads or tails—this is a dichotomy, a nominal measure with only two categories. Table 3.1 listed the results of spinning a coin 500 times. Taking basic probability theory as our guide, we would expect to see 250 heads and 250 tails—a *binomial distribution* of 0.5 (50%) for each—if the coins were not biased. In fact, the observed distribution was 253 heads (50.4%) and 247 tails (49.6%), which is a small departure from the expected *binomial distribution*. This is

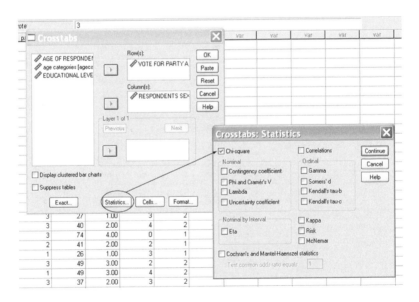

Figure 10.11: Setting the x^2 statistic

unlikely to be statistically significant, but that would need to be tested.

As a more practical example of the use of the *binomial test* in the social sciences, consider the issue of gender balance in the labour market. Equal opportunities good practice monitoring dictates that a company ought to be more pro-active in ensuring its workforce reflects a gender balance typical of the labour market rather than simply focus on good selection processes. The key question is: to what extent is the company monitoring gender balance and attempting to address any imbalance? Clearly, before that question can be answered, the company needs to determine if it has a gender imbalance in its employees.

Gender is a dichotomous variable (male and female) and the *binomial test* is used to determine statistical significance— i.e. is any observed difference in the employee gender balance significantly different from that of the labour market generally? The *binomial test* will compare the proportion of each gender in the sample (i.e. this company's employees) with a realistic comparative value. Whereas when spinning a coin the expected proportions will be 50/50, that will not always be the case worth binomial distributions. In this case, we want to use the known distribution of the genders in the labour market against which to test the company's claims to be a good employer in this respect. The UK Labour Force Survey will provide that information, and since the pattern may vary from region to region, the researcher feels it only fair to the company to make comparisons with the local labour market. Government data shows that 51.4% of those who are economically active (i.e. in work or seeking it) are male, and 48.6% are female.

However, the database available to the researcher uses the codes m and f for male and female respectively, and the binomial test requires data to be numeric. The gender variable needs to be recoded before the test can be applied.

Hypothesis: Company A claims it employs higher proportion of female staff than the labour market generally, and this is an indication of its positive action policies.

Testing the hypothesis in SPSS

1. The **gender** variable needs to be recoded into numeric values. As with the salary data earlier in the chapter, the safest thing is to recode values into a new variable, thereby preserving the integrity of the original data. Select **Transform** > **Recode** > **Into Different Variable**. Create the new variable as before and click *Change*, followed by clicking *Old and New Values*. Recode the old value into the new value and add it to the variable list as shown in fig 10.12. Click Continue to return to the Recode dialog box. Click OK to run the recode (fig 10.13).

2. Run the *binomial test* from **Analyze** > **Non-parametric tests** > **Binomial**. Transfer the test variable to the Test Variable List box (fig 10.14) and set the Test Proportion. The default value of 0.50 will already be in there (based on the assumption of equal proportions) but you can type in the test proportion you require. This is 0.514 (.514 in the figure) because this is the proportion of economically active males in the labour market. Because males are coded 1 in the database, it is that proportion that needs to be in the box. If females had been coded 1, then 0.486 would be the test statistic. Click OK to run the analysis.

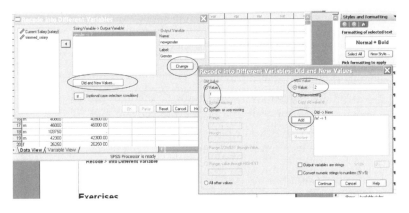

Figure 10.12: Recoding values into a new variable

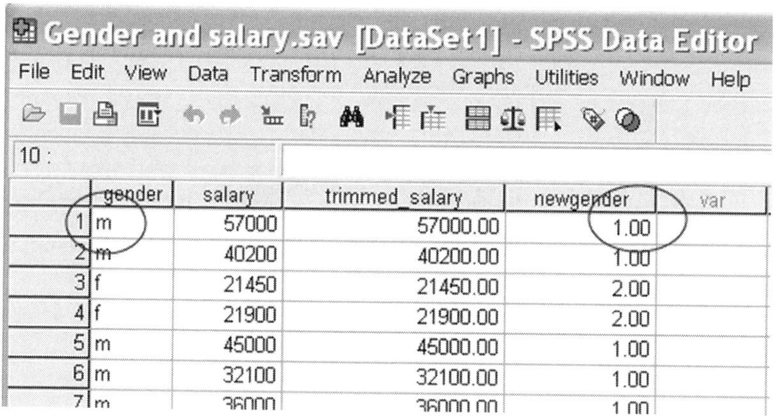

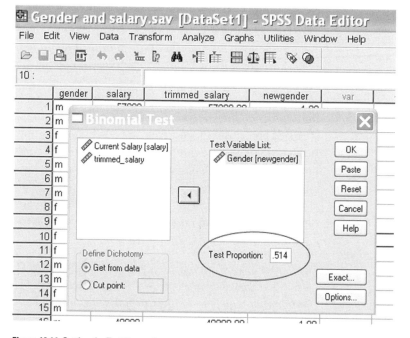

Figure 10.13: The new variable recodes the alphabetic format of gender to a numeric code – SPSS works better with numbers

Figure 10.14: Setting the Test Proportion

3. Table 10.13 shows the result. There is actually a slightly higher proportion of male employees but this is not statistically significant (p>0.05).

Binomial test						
		Category	N	Observed prop.	Test prop.	Asymp. sig. (1-tailed)
Gender	Group 1	male	258	.544	.514	.101(a)
	Group 2	female	216	.456		
	Total		474	1.000		

ª Based on Z Approximation

Conclusion: The company's claims are not substantiated. There is no statistically significant difference between the gender balance in its workforce compared to the general labour market in the region.

Mann-Whitney U test

This test, and the *Wilcoxon Matched-Pairs Signed Ranks Test* are the equivalent of the t-test for ordinal dependent variables. In the case of *Mann-Whitney* the test is used with independent samples. It is based on ranking responses of responses to the dependent ordinal variable and compares the number of times a score from one of the sample groups is ranked higher than a score from the other sample group. If the two samples are similar, we would expect the scores on the ordinal scale to be similar and so the mean ranking (i.e. the average position in the ranking) should be similar also. The *Mann-Whitney U test* determines whether any difference in the mean ranking is statistically significant.

Consider the following hypothesis:

Hypothesis: In a survey of attitudes and values, the researcher believes that female respondents have a significantly higher level of trust in the National Health Service than male respondents.

Testing the hypothesis in SPSS

1. The *Mann-Whitney test* is accessed from the **Analyze > Non-parametric tests > 2 Independent Samples**. In the dialog box that opens (fig 10.15), move the dependent variable (trust in the NHS) to the Test Variables box and gender (the independent variable) to the Grouping Variable box. Make sure the Mann Whitney U checkbox is ticked. Click Define groups to complete the group codes (i.e. 1 for male and 2 for female). Click Continue then OK to run the test.

2. The results show a highly statistically significant result, but not what was expected by the original hypothesis. The Ranks table shows that male respondents are on average ranked higher in their responses than females. N represents the number of cases in the analysis and the Sum of Ranks is the product of N and Mean Rank. The Test Statistics table (Table 10.14) shows that this result is highly significant at p<0.001. However, this is given a two-tailed result, and as this is a directional hypothesis, we need to halve that value, but it is still highly significant.

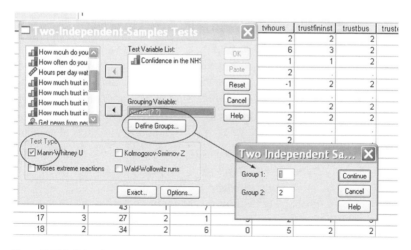

Figure 10.15: Defining the groups for the t-test

Ranks				
	Gender	N	Mean rank	Sum of ranks
Confidence in the NHS	Male	432	458.36	198011.00
	Female	556	522.58	290555.00
	Total	988		

Test statistics(a)	
	Confidence in the NHS
Mann-Whitney U	104483.000
Wilcoxon W	198011.000
Z	–3.885
Asymp. Sig. (2-tailed)	.000
a Grouping Variable: Gender	

Conslusion: The original hypothesis has to be amended to state that male respondents tend to trust the NHS more than female respondents and this result is highly statistically significant at p<0.001.

Wilcoxon matched pairs signed rank test

Like *Mann-Whitney*, this test ranks the scores from each sample group and compares them. If there is no difference between the groups then the mean ranking should be the same. This test is applied when exploring the relationship between an ordinal scale for related samples. For example, you may want to see if the opinions of attendees at a public meeting with a prospective Parliamentary election candidate have been changed by listening to the candidate. This is an example of where we would use a two-tailed value for probability because we cannot be sure what the impact will be. We are also using this study to develop a hypothesis rather than test one. Attendees are asked if they like, dislike or are neutral about the candidate before the debate, and are asked again (completion of a card that is handed in and matched to the pre-meeting response later) following the debate to see if they have changed their minds and in which direction.

> Question: What is the impact of the public debate with the candidate on their views of him?

Using SPSS to develop a hypothesis

1. Set up the *Wilcoxon matched pairs signed rank test* from **Analyze > Nonparametric Tests > 2 Related Samples**. In the dialog box that opens select the two related variables (in this case Opinion1 and Opinion2) and transfer them to the Test Pair(s) list. Make sure the Wilcoxon checkbox is ticked (fig 10.16). Click OK to run the test.

2. The SPSS printout is interpreted in the following way. Ties refer to the cases where there was no change. 72 attendees showed a less favourable opinion after the meeting, while 71 shows a more favourable opinion. The overwhelming

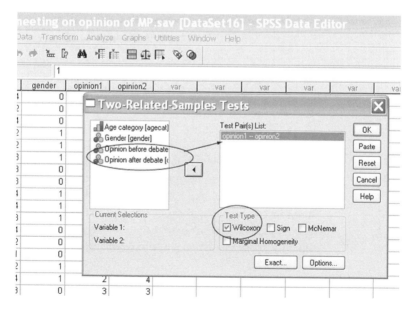

Figure 10.16: Setting the test type

result seems to be that the meeting had made little difference, and the improved opinions are neutralized by the same number (almost) of negative changes in opinion. It should come as no surprise then, that the Test Statistics show no statistical significance ($p = 0.907$).

Ranks				
		N	Mean rank	Sum of ranks
Opinion after debate–Opinion before debate	Negative Ranks	72(a)	72.28	5204.00
	Positive Ranks	71(b)	71.72	5092.00
	Ties	847(c)		
	Total	990		

[a] *Opinion after debate < Opinion before debate*
[b] *Opinion after debate > Opinion before debate*
[c] *Opinion after debate = Opinion before debate*

Test Statistics(b)	
	Opinion after debate– Opinion before debate
Z	–.117(a)
Asymp. Sig. (2-tailed)	.907

[a] *Based on positive ranks*
[b] *Wilcoxon Signed Ranks Test*

Conclusion: The public meeting has not resulted in any statistically significant change in opinion of the candidate as a direct result of the meeting. Indeed, the evidence suggests that opinions were quite firmly established prior to the meeting with relatively little movement (i.e. only 143 out of 990 respondents) in either direction.

Exercises

Task 1

The following is a list of data files used in this chapter and which can be downloaded from the Studymates website. They are in Excel format and can be imported into any version of SPSS:

- Attitudes and values
- Gender and salary
- Gender v time to do a task
- Impact of training
- Voting pattern

Download these files and practise the exercises described in the chapter to familiarize yourself with the techniques.

Task 2

If you haven't already done so, download the file on student satisfaction (satisfaction.xls and its coding frame) and answer the following research questions using SPSS:

1. Is there any difference in the overall satisfaction levels of:
   a. Access versus undergraduate students
   b. Male students versus female students
2. Are there any particular groups of students who felt more valued by tutors than other groups?
3. Are there any particular groups of students who felt more challenged by their learning experience than other groups?

11

Correlation: Exploring the strength of a relationship

Learning outcomes:

It is one thing identifying and quantifying the relationship between two variables, and determining that it is a statistically significant relationship, it is quite another to say how strong that relationship is. Yet another problem is that of determining the extent to which an independent variable (such as gender) has a real causal effect on a dependent variable (such as employment status). Chapter 9 explored this in some detail, looking at the different types of relationships—spurious, intervening and interactive variables. Correlation is a method of exploring the strength of a relationship between variables. By the end of this chapter you should:

■ Understand the principle of 'correlation' as the strength of association between variables;

■ Know that the 'correlation co-efficient' is a measure of the strength of the association;

■ Understand that the co-efficient may vary between −1 (perfect inverse association), through 0 (no correlation) to +1 (perfect direct correlation);

■ Be able to apply and calculate appropriate correlation tests (i.e. *Pearson's r, Spearman's rho, phi*).

Understanding correlation

Fig 11.1 illustrates the basic principle of the bivariate relationship discussed in chapter 10. As the independent variable varies, so does the dependent variable. In other words there is an apparent associative or causal relationship.

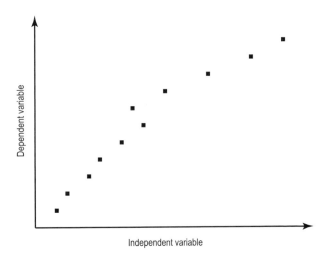

Figure 11.1: Positive relationship between variables

As has already been explained in chapter 9, care must be exercised in making any claims of causality because there may be factors that are either unknown to the researcher, or simply have not been measured and incorporated into the analysis.

The association shown in fig 11.1 is a positive one—i.e. as the independent variable varies, the dependent varies in the same direction. Fig 11.2 shows a negative association. As the independent variable varies, the dependent variable also varies, but in the opposite direction.

The strength of this relationship is called *correlation* and is measured using the *correlation coefficient* (symbolized by *r*). The degree of association between two variables is quite important in attempting to describe how close the link is between the two. The idea of *correlation* helps us identify the degree to which one variable influences another. Note that this is a two-way interaction: if variable x varies as variable y varies; variable y varies as variable x varies.

Fig 11.3 illustrates the quantification of *correlation*. (a) shows the distribution of cases plotted on a graph. They are clustered together very closely around a straight line showing that there is a very strong relationship between the values

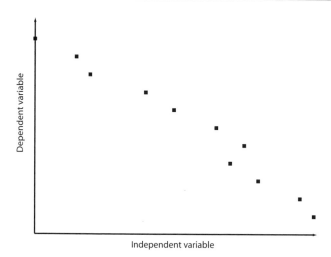

Figure 11.2: Negative relationship between variables

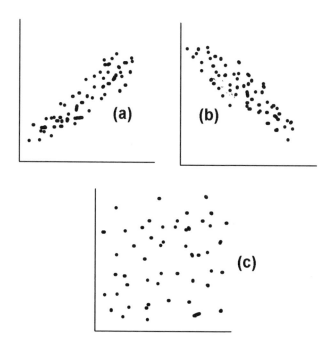

Figure 11.3: Quantifying correlation

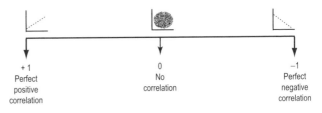

+ 1
Perfect
positive
correlation

0
No
correlation

−1
Perfect
negative
correlation

Figure 11.4: The correlation coefficient and effect on data plots

on one variable with the value on another. (b) also shows a strong correlation, but it is in the negative direction (i.e. as the independent variable increases, the dependent variable decreases). (c) shows no correlation whatsoever as the plots are scattered all over the place without any pattern to their distribution.

The *Correlation coefficent* is represented numerically as a decimal ranging from −1 to +1. This is called the *correlation coefficient.* +1 and −1 represent perfect *correlation,* while zero means there is no *correlation* whatsoever (fig 11.4).

Using the correlation coefficient

For example, a correlation of +0.6 is not a perfect relationship, but it is still positive as both variables will increase with each other. Being less than 1 means there are other factors that are affecting the relationship.

It is rare that any one variable is totally dependent on another, as has been discussed already in chapter 9, and other factors have an effect that moderate the relationship between any two variables, so the relationship is less than perfect. For example, the correlation coefficient may be +0.6—still positive (i.e. both variables increase with each other). Taken together with *statistical significance,* correlation is a key measure that helps us understand the strength of the relationship between two variables. The *coefficient of determination* helps us assess the extent to which two variables may have a causal relationship (see below).

It is worth spending a moment or two looking at hypothetical examples to illustrate what this means in real life.

If you were to examine the rate of spending on primary phase education and the effect this had on the SATs results at age eleven, you would look at the *correlation* between the two (as well as its statistical significance). It is likely that this will show a positive correlation, but this is unlikely to be perfect. A *correlation coefficient* of +0.6 would be quite strong in view of the other likely social and economic factors that will play their part. A *correlation coefficient* of +0.1 is very low and will make you question the wisdom of throwing so much money at the problem, because clearly there must be other more important factors.

Another example to illustrate *negative correlation* might be the effect of increasing investment in supporting their local community in setting up and maintaining Neighbourhood Watch schemes. A test of this investment would be the effect on property related crime. Here, you would be looking for a strong *negative correlation* As investment was increased, you would expect property related crime to decrease. Again, you would be looking for a strong negative value (e.g. –0.7).

It is possible to talk about weak, moderate and strong correlations. A rule of thumb adopted is that positive or negative correlations below 0.2 are low, between 0.2 and 0.40 is , 0.4 to 0.7 is , 0.7 to 0.9 is , and anything above this is high.

Coefficient of determination (R^2)

The *coefficient of determination* tells us how much of the variation in the dependent variable can be explained by the variation in the independent variable. R^2 is calculated by squaring the *correlation coefftient* and expressing the resultant decimal as a percentage. Table 11.1 shows how the *coefficient of determination* is related to the *coefficient correlation*

Correlation coefficient	Coefficient of determination
0.1	1% (0.01)
0.2	4% (0.04)
0.3	9% (0.09)
0.4	16% (0.16)
0.5	25% (0.25)
0.6	36% (0.36)
0.7	49% (0.49)
0.8	64% (0.64)

Table 11.1:
Coefficients of determination

(Continued)

Correlation coefficient	Coefficient of determination
0.9	81% (0.81)
1.0	100% (1.0)

What the table clearly shows is that the correlation needs to be very high before more than half the variation in the dependent variable can be explained by the independent variable. For example, a correlation coefficient of 0.7 means that only 49% of the variation can be explained by the independent variable; 51% is explained by other factors (variables). This is why the researcher has to be cautious about making any claims relating to causality. The fact that two variables are linked in this way, and with statistically significance, is not sufficient to say the two have a direct causal relationship. More direct evidence will be needed.

Parametric and non-parametric coefficients

There are a number of forms of *correlation coefficients* developed by statisticians that are designed for use with interval, ordinal and nominal variables. SPSS will calculate these, but you also need to determine statistical significance as well otherwise the *correlation coefficient* is meaningless (i.e. because the observations are likely to be the result of chance). The three most useful ones are described here.

Pearson's product-moment correlation (Pearson's *r*)

This is designed to be used with interval variables. *Pearson's r* makes the following assumptions about the variables:

- Both variables are interval
- The relationship between the two variables is a linear one

Q: In the study to evaluate the impact of training on the time to complete a task, the company also wants to know if there is a relationship between the time before training and the time after training.

Testing the correlation in SPSS

1. Set up the correlations analysis from **Analyze** > **Correlate** > **Bivariate** (fig 11.5).

2. In the Bivariate Correlations dialog box (fig 11.6), move the variables to be tested to the Variables box. The specific correlation coefficient needs to be selected as does the test of significance. Note that the Flag significant correlations checkbox is also ticked. Click OK to run the analysis.

3. The resulting table (Table 11.2) is a *correlation matrix* that shows the correlation coefficient between the two variables (clearly *r* between the same variable will always be 1).

   a. There is a moderate positive correlation between the times before and after training and this is statistically significant. The 2-tailed significance is $p < 0.01$, so even taking a 1-tailed result, the correlation is significant.

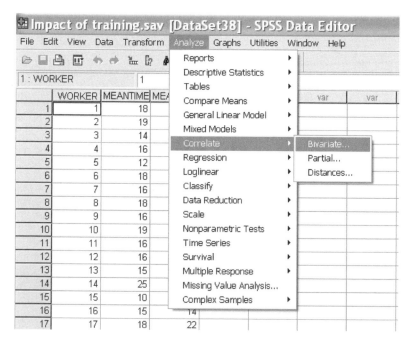

Figure 11.5: The menu route to bivariate correlation tools and statistics

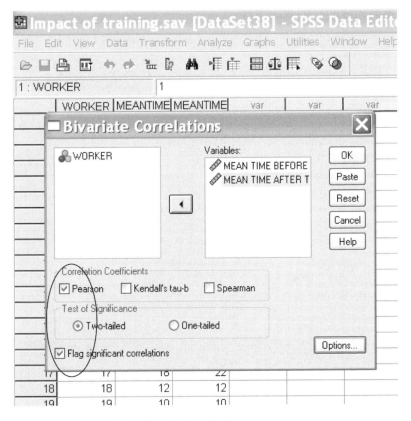

Figure 11.6: Selecting the correlation statistic required

b. The coefficient of determination is 0.32–32% of the variation is the result of the relationship between these two variables.

Table 11.2:
Correlation matrix

Correlations		Mean time before training	Mean time after training
Mean time before training	Pearson Correlation	1	.563(**)
	Sig. (2-tailed)		.001
	N	30	30

(Continued)

Correlations		Mean time before training	Mean time after training
Mean time after training	Pearson Correlation	.563(**)	1
	Sig. (2-tailed)	.001	
	N	30	30

◀ Table 11.2: (Continued)

** Correlation is significant at the 0.01 level (2-tailed).

Conclusion: There is a moderate correlation between the time taken to perform the task before and after the training. Since this is positive, it means that even with training, if the time taken by a worker before training is higher than that of other workers, it is more likely that their time will still be higher, although probably shorter than before training—see the t-test of this data in chapter 10).

Spearman's rho

A different measure of *correlation* is recommended for ordinal measures. This is called *Spearman's rho*. It makes fewer assumptions about the data and works by comparing the rankings (order) of the variable values rather than comparing the interval differences.

In SPSS, you can specify *Spearman's rho* when you ask the program to set up a bivariate correlation matrix. The result is interpreted in much the same way as for *Pearson's r* because the statistic means the same thing to all intents and purposes.

Hypothesis: There is a statistically strong correlation between people's confidence in the National Health Service and their confidence in the education system.

Testing the correlation in SPSS

1. If this hypothesis is correct, the result should show significant positive correlation and preferably a high level of correlation—so that the *coefficient of determination* is also high. From **Analyze** > **Correlate** > **Bivariate**, transfer the appropriate variables to the Variables box, and select *Spearman* as the correlation statistic (deselect *Pearson* if

this is checked). As this is a directional hypothesis, select the one-tailed test of significance. The resulting table is shown in Table 11.3.

Table 11.3:
The SPSS Output
table for the
Spearman's rho
correlation test

Correlations			Confidence in the NHS	How much trust in education system
Spearman's rho	Confidence in the NHS	Correlation Coefficient	1.000	.299(**)
		Sig. (1-tailed)	.	.000
		N	988	980
	How much trust in education system	Correlation Coefficient	.299(**)	1.000
		Sig. (1-tailed)	.000	.
		N	980	992

** Correlation is significant at the 0.01 level (1-tailed).

Conclusion: There is a weak, but statistically significant correlation between respondents' confidence in the NHS and trust in the educational system. The coefficient of determination is 8.9% so while there is a link between the two attitudes it is unlikely to be a causal relationship. It is much more likely that there are other factors that are influencing both attitudes (i.e. this may be a spurious relationship—see chapter 9).

Phi

The value of the *correlation coefficient* is really felt with interval data. After all, it is a measure of the relationship between values that are changing. *Spearman's rho* does address the need to examine this relationship with ordinal values. There is also a measure that looks at the strength of the relationship between dichotomous nominal variables (e.g. Gender, Yes/

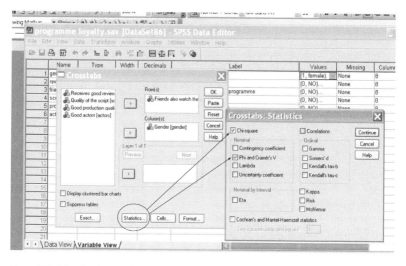

Figure 11.7: Selecting the correlation statistics for dichotomous variables

No, etc). This is called *Phi* and SPSS will calculate *Phi* as part of the **Crosstabs** command in the **Statistics** menu.

Hypothesis: In a survey of a sample of viewers of a popular television series, the production company wants to know some of the factors that help to maintain viewer loyalty. A particular interest is the effect of gender on loyalty. The perception amongst the production team is that a key relationship is the fact that female viewers are influenced by the fact that their friends also watch the programme.

Testing the correlation in SPSS

1. The variables in the programme loyalty survey are all dichotomous nominal measures. The phi statistic is generated from **Analyze > Descriptive Statistics > Crosstabs**. In the dialog box that opens, transfer the dependent variable (in this case, friends) to the Rows box, and the independent variable (in this case, gender) to the Columns box. Click Statistics to select phi and chi-square (fig 11.7). Click Continue then Ok to run the analysis.

2. The three important tables are shown in Table 11.4. The crosstabulations table shows the distributions of responses to the question (a very clear difference between males and females) and the Chi-Square Tests table shows that this distribution is very highly significant (even at the one-tailed test level). Phi also shows the correlation between gender and the friends variable to be fairly high (the *coefficient of determination* is 61%). The correlation is shown as negative because males are coded 2 and females coded 1. See 'Symmetric measures table' in Table 11.4.

▶
Table 11.4:
The SPSS
Output table for
Phi—along with
Chi-square test
data

Friends also watch the programme * Gender crosstabulation

			Gender		
			female	male	Total
Friends also watch the programme	NO	Count	31	396	427
		% within Gender	7.0%	85.2%	47.1%
	YES	Count	410	69	479
		% within Gender	93.0%	14.8%	52.9%
Total		Count	441	465	906
		% within Gender	100.0%	100.0%	100.0%

Chi-square tests

	Value	df	Asymp. Sig. (2-sided)	Exact Sig. (2-sided)	Exact Sig. (1-sided)
Pearson Chi-Square	554.514[b]	1	.000		
Continuity Correction[a]	551.382	1	.000		
Likelihood Ratio	638.107	1	.000		

(Continued)

Chi-square tests					
	Value	df	Asymp. Sig. (2-sided)	Exact Sig. (2-sided)	Exact Sig. (1-sided)
Fisher's Exact Test				.000	.000
Linear-by-Linear Association	553.901	1	.000		
N of Valid Cases	906				

◄ Table 11.4: (Continued)

ª Computed only for a 2 x 2 table.
ᵇ 0 cells (.0%) have expected count less than 5. The minimum expected count is 207.84.

Symmetric measures			
		Value	Approx. Sig.
Nominal by Nominal	Phi	–.782	.000
	Cramer's V	.782	.000
N of Valid Cases		906	

ª Not assuming the null hypothesis.
ᵇ Using the asymptotic standard error assuming the null hypothesis.

Conclusion: The production team's observation that female viewers loyalty is influenced partly by the fact that their friends also watch the programme is supported by the data which shows a statistically significant correlation.

Selecting the best measure

So far, the descriptions of *correlation coefficients* assume that your variables are of the same type. This is not always the case. You may need to compare interval with ordinal, or ordinal with dichotomous, or interval with dichotomous. You should adopt the measure that relates to the lowest level of measure. So, for example, if you are comparing interval with ordinal, you will probably use *Spearman's rho*.

Exercises

1. Why is it useful to understand the strength of the relationship between two variables using correlation?
2. If the correlation coefficient between the two variables of age and reaction times in an experiment to measure responses to an emergency is −0.45 and the one-tailed test of significance shows $p < 0.01$.
   a. What does the negative sign mean?
   b. What is the *coefficient of determination* and what does it mean?
   c. Provide an interpretation of the results.

Assignment 1

Use the data files that are downloadable from the Studymates website and explore the correlations between pairs of variables using SPSS.

Assignment 2

You have been asked to carry out a piece of research to test the idea that people's ability to memorize a list of words is influenced by their age—the ability to memorize in a given time reduces with increasing age. Develop a research hypothesis and design an experiment to test the hypothesis. Include in the design:

a. The two variables you will collect data on and the nature of the data (nominal, ordinal or interval);
b. The statistical tests of significance and correlation you will use;
c. If you have the time and resources to do so, carry out the experiment and analyse your results.

Index